SpringerBriefs in Electrical and Computer Engineering

Series Editors

Woon-Seng Gan, School of Electrical and Electronic Engineering, Nanyang Technological University, Singapore, Singapore

C.-C. Jay Kuo, University of Southern California, Los Angeles, CA, USA

Thomas Fang Zheng, Research Institute of Information Technology, Tsinghua University, Beijing, China

Mauro Barni, Department of Information Engineering and Mathematics, University of Siena, Siena, Italy

SpringerBriefs present concise summaries of cutting-edge research and practical applications across a wide spectrum of fields. Featuring compact volumes of 50 to 125 pages, the series covers a range of content from professional to academic. Typical topics might include: timely report of state-of-the art analytical techniques, a bridge between new research results, as published in journal articles, and a contextual literature review, a snapshot of a hot or emerging topic, an in-depth case study or clinical example and a presentation of core concepts that students must understand in order to make independent contributions.

More information about this series at http://www.springer.com/series/10059

Xiaoming Chen · Qiao Qi

Convergence of Energy, Communication and Computation in B5G Cellular Internet of Things

Springer

Xiaoming Chen
Zhejiang University
Hangzhou, Zhejiang, China

Qiao Qi
Zhejiang University
Hangzhou, Zhejiang, China

ISSN 2191-8112 ISSN 2191-8120 (electronic)
SpringerBriefs in Electrical and Computer Engineering
ISBN 978-981-15-4139-1 ISBN 978-981-15-4140-7 (eBook)
https://doi.org/10.1007/978-981-15-4140-7

This Springer imprint is published by the registered company Springer Nature Singapore Pte Ltd.
The registered company address is: 152 Beach Road, #21-01/04 Gateway East, Singapore 189721, Singapore

Preface

With the explosive growth of Internet of Things (IoT), a massive number of IoT devices desire to access wireless networks for realizing various advanced applications, e.g., smart city, industry automation, and remote medicine. It is predicted that over 75.4 billion devices will be linked to the Internet all over the world by 2025. Although many IoT devices can be served by short-range radio technologies typically applicable for indoor environments, such as WiFi, ZigBee, and Bluetooth, a significant proportion of IoT devices have to be enabled by wide-area networks. To this end, cellular IoT is emerging as a promising solution, which can interconnect low power, massive connectivity, and wide coverage IoT devices at a low cost. In 2015, 3GPP has identified cellular IoT as one of the main use cases of 5G wireless networks, and issued a specification for cellular IoT in Release 13.

With 5G and even beyond wireless networks, cellular IoT can further unlock the potential of smart devices. As a result, a movie can be downloaded with lightning speeds, autonomous vehicles can be safer due to faster reaction times, and industries can be revolutionized with smart machinery and stock. Despite cellular IoT having the characteristics of low power, massive connectivity, and wide coverage, there exists an unprecedented pressure on the backhaul link for data processing at cloud servers. In order to support real-time processing of mass data from terminal devices, future cellular IoT has to be a large-scale edge-intelligent network. For example, the video mentors in the street are not only a communication node, but also a computation node. By exploiting the potential of a massive number of IoT devices, it is possible to realize high edge intelligence.

Without a doubt, the key to edge intelligence lies in efficient computation and communication with a massive number of IoT devices. However, the small battery capacity heavily limits the functions of these edge devices. In this context, this book dedicates to investigate the convergence of energy, communication and computation in beyond 5G (B5G) cellular IoT. Both theory and technique have been addressed, with more weight placed on the key techniques. In Chap. 1, we introduce the characteristics of B5G cellular IoT and its key techniques for realizing effective convergence of energy, communication and computation. Next, Chap. 2 addresses the issue of convergence of energy and communication in B5G cellular IoT with a massive number of devices enabled by simultaneous wireless information and power transfer. In Chap. 3, we consider a wireless powered computation-centric B5G cellular IoT network, and provide an effective solution for the convergence of energy and computation. Then, Chap. 4 investigates the issue of convergence of communication and computation in B5G cellular IoT, and a new framework integrating communication and computation is proposed and optimized. Furthermore, a sustainable B5G cellular IoT integrating energy, communication and computation is discussed in Chap. 5, and a beamforming algorithm is designed to improve the overall performance. Finally, we make a summary about the convergence of energy, communication and computation in B5G cellular IoT, and point out the future research directions for cellular IoT with high edge intelligence in Chap. 6. It is sincerely expected that this book can provide useful insights for the analysis, design and optimization of B5G cellular IoT.

Hangzhou, China
February 2020

Xiaoming Chen
Qiao Qi

Contents

Chapter 1
Introduction

Abstract In this chapter, we first introduce the cellular IoT, which utilizes existing cellular networks that we are using every day in current human-centric communication (HCC), e.g., audio and video, to provide massive machine-type communications (mMTC), e.g., sensing and monitoring. Then, we present the development of cellular IoT based on 3GPP releases, and discuss the prospect of cellular IoT in B5G wireless networks. Next, we give an overview of energy, communication and computation in B5G cellular IoT, and introduce the relevant techniques for realizing their convergence in B5G cellular IoT. Finally, we present the objective and content of this book.

1.1 Cellular IoT

With the rapid development of the IoT, the number of wireless devices continues to surge, and relevant applications emerge one after another. The era of internet of everything is coming, promoting smart industries across energy, transportation, healthcare, etc., and revolutionizing the way people live and work [1–3]. The possibilities are endless when it comes to IoT and statistics definitely reflect that in Fig. 1.1. It is predicted that there will be more than 75 billion IoT devices worldwide by 2025, a fivefold increase in ten years [4].

In order to unlock the potential of IoT, wireless devices have to be connected by using wireless communications technologies. Although many IoT devices can be served by short-range radio technologies typically applicable for indoor environments, such as WiFi, ZigBee, and Bluetooth, a large proportion of IoT devices have to be enabled by wide-area networks (WANs). Low-power WAN (LPWAN) is a wireless communication technology that interconnects low-bandwidth, battery-powered devices with low bit rates over long ranges [5]. Currently, there are two connectivity tracks for LPWAN, one operating on an unlicensed spectrum such as SigFox and LoRa [6, 7], and the other operating on a licensed spectrum such as cellular IoT [8, 9]. The comparison of different wireless communication technologies for IoT networks is shown in Table 1.1.

X. Chen and Q. Qi, *Convergence of Energy, Communication and Computation in B5G Cellular Internet of Things*, SpringerBriefs in Electrical and Computer Engineering, https://doi.org/10.1007/978-981-15-4140-7_1

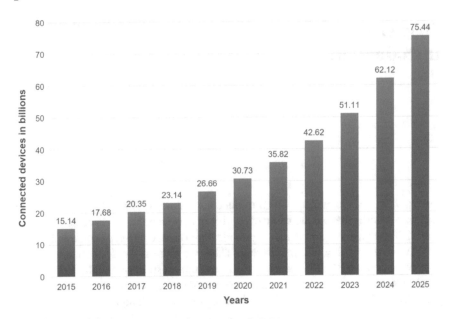

Fig. 1.1 Number of IoT connected devices from 2015–2025

Table 1.1 Comparison of wireless communication techniques for IoT networks

	Bluetooth	Zigbee	WiFi	LoRa	SigFox	Cellular IoT
Spectrum	Unlicensed	Unlicensed	Unlicensed	Unlicensed	Unlicensed	Licensed
Connectivity	Small	Medium	Large	Massive	Massive	Massive
Range	Short	Short	Medium	Long	Long	Long
Power	Medium	Medium	High	Low	Low	Medium
Delay	Short	Short	Short	Short	Short	Short
Security	Low	Medium	Medium	Medium	Medium	High
Mobility	Not	Not	Not	Yes	Yes	Yes
Cost	Low	Low	Low	High	Medium	Low

1.1.1   What is Cellular IoT?

Here, you might wonder "What is cellular IoT?", but I am sure that you are familiar
with the underlying technology. Do you know how your smartphone can connect
the Internet? It accesses a base station (BS) that is broadcasting over a specific
area, namely cell, c.f., Fig. 1.2. This is why mobile phones are sometimes called
"cellphones". When you realize that your mobile phone uses cellular networks to
connect to internet, cellular IoT gets a lot easier to understand. It is the wireless

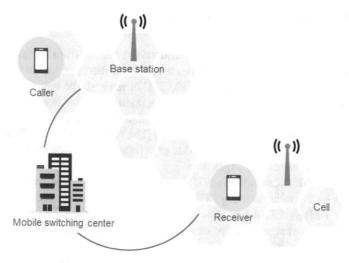

Fig. 1.2 The access of mobile phones

communication technology for an IoT device to transfer its data to a WAN. Cellular IoT is tailored for IoT devices and applications, and there are three main reasons as follows:

(1) **Coverage**: Cellular networks are widespread and ubiquitous, covering 90% of the world's population. Other wireless communication technologies like WiFi do not have the same scale, requiring users to search for and connect to a local network. With the integration of satellite communication in B5G wireless networks, B5G cellular IoT even can provide a coverage in non-man's land, e.g., desert, forest, and ocean.

(2) **Range**: As mentioned above, short-range wireless communication technologies like WiFi have an advantage in high-bandwidth communications, but are not available to long-distance communications. Cellular IoT meets the needs of low-power devices distributed in a very large range. With the development of B5G wireless networks, cellular IoT can even provide a high rate similar to WiFi.

(3) **Feasibility**: Mobile operators, RF providers, and wireless infrastructure companies have invested heavily in cellular networks to provide secure, reliable services to as many customers as possible. By leveraging existing infrastructure and mature technology, it is feasible to utilize cellular IoT for connecting billions of IoT devices with little additional investment.

Due to the wide coverage, flexible resource managements and so on, cellular networks have become key technologies to support the massive IoT connections. Especially with B5G wireless networks on the horizon, the future looks great for cellular IoT [10, 11].

1.1.2 Development of Cellular IoT

In recent years, cellular IoT has gained considerable attention from academia, industries, regulators and standardization bodies to enable the incorporation of IoT devices in the existing cellular infrastructures [12, 13]. In general, the cellular IoT refers to 3GPP-supported cellular communication technologies with licensed spectrum, including extended coverage-global system for mobile communication (EC-GSM) [14], long-term evolution for machine (LTE-M) [15], narrowband IoT (NB-IoT) [16], etc. For intuitive observation, the evolution of cellular communication technologies for IoT networks is shown in Fig. 1.3.

In 2013, LTE-M was issued in 3GPP R12, aiming to meet the requirements of IoT devices based on the existing LTE networks with an upstream and downstream data rate of 1 Mbps. LTE-M has four basic advantages of the LPWAN technology including wide coverage, massive connections, low power consumption and low module costs. In specific, a cell of the LTE-M network can support almost 100,000 connections, and LTE-M terminals can stand by for up to 10 years. Compared with the existing technologies under the same licensed spectrum (700–900 MHz), LTE-M achieves a transmission gain of 15 dB, which boosts the coverage of LTE networks remarkably.

In 2014, EC-GSM was put forward in the research project of 3GPP GSM edge radio access network (RAN). In comparison to traditional GPRS, a wider coverage was achieved with 20 dB, and five major goals were proposed, namely improve indoor coverage, support massive connectivity, simplify the devices, lower power consumption, and reduce latency. However, with the continuous development of the technology, cellular IoT should be renewed and upgraded to meet the higher demands, resulting in the emergence of NB-IoT.

In 2015, the 3GPP RAN initiated the research on a new air-interface technology called clean slate cellular IoT for narrowband wireless access, including the narrowband cellular IoT (NB-CIoT) and narrowband LTE (NB-LTE). These two technologies are incompatible with each other, but compatible with the existing LTE. Then in 2016, in order to obtain a unified solution, NB-IoT as a combination of NB-CIoT and NB-LTE was born in 3GPP R13. As a leading LPWAN technology, NB-IoT enables low-complexity devices at a low-cost with long battery life (more than 10 years), deployed in massive numbers (over 50,000 connections per cell) and with an extended geographical cell range (even up to 100 km).

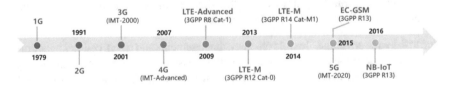

Fig. 1.3 The evolution of cellular communication technologies for IoT networks

Beyond 3GPP Release 13, there is a development roadmap of cellular IoT technology with further enhancements to meet massive IoT connectivity needs in the future [17]. For example, Release 14 will bring new capabilities such as single-cell multicast to both eMTC and NB-IoT, enabling easy over-the-air firmware upgrades as well as enhanced device positioning for asset location tracking. In addition, there are two ongoing Release 15 work items that further enhance eMTC and NB-IoT, including new features such as wake-up receiver and TDD support for NB-IoT.

1.1.3 Cellular IoT in Beyond 5G Networks

As shown in Fig. 1.4, 5G is envisioned to support a multitude of service and devices, thus it needs to be adaptable to a huge variance of requirements around coverage, throughput, capacity, latency, and reliability [18–21]. With B5G wireless networks, IoT will no longer be constrained by network resources, unlocking the potential of smart devices. As a result, downloads can take place at lightning speeds, autonomous vehicles can be safer thanks to faster reaction times, and industries can be transformed with new ways to connect machinery and stock.

In general, 5G New Radio (NR) networks can support three main use cases, c.f. Fig. 1.5, namely enhanced mobile broadband (eMBB), ultra-reliable and low latency communications (uRLLC), and massive machine-stype communications (mMTC) [22].

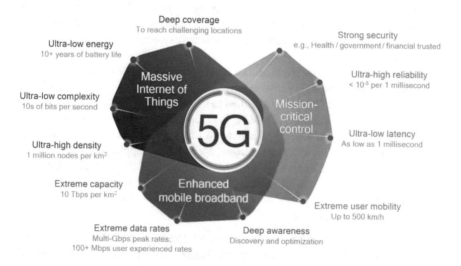

Fig. 1.4 The key performance indicators of 5G wireless network

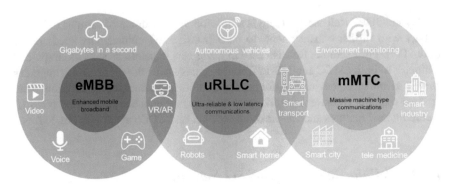

Fig. 1.5 Three main use cases of 5G wireless networks

(1) eMBB is a natural unfolding of LTE, whose primary goal is to increase the user data rates and network spectral efficiency. The enhancements provided by eMBB target mainly human-type traffic, such as high-speed wireless broadband access, ultra-high quality video streaming, virtual reality (VR) and augmented reality (AR).

(2) uRLLC, also known as mission-critical IoT, envisions transmission of moderately small data packets (in the order of tens of bytes) with extremely high-reliability, ranging between 99.999 and 99.9999999%, i.e. down to 10^{-9} packet error probability. The user plane latency requirement is most commonly defined to be 1 ms, including uplink (UL) and downlink (DL) roundtrip transmission. uRLLC is seen as an enabler of safety systems, wireless industrial robots, autonomous vehicles (cars, trucks, drones), immersive VR with haptic feedback, tactile Internet, and many others which may not even be foreseen at this point.

(3) mMTC aims to provide service to a massive number of devices, out of which only a certain fraction are active at a given time. The packet lengths in mMTC are comparable to uRLLC, being assumed to be rather short, in the range of tens of bytes. The main challenge of mMTC is to enable access for sporadically active devices, such that at any given instant an unknown subset out of the massive set of devices wishes to send messages. Most mMTC applications do not have strict delay requirements. Any massive deployment of connected devices falls into the category of mMTC, where devices are mainly sensors used to gather measurements from various environments, such as weather, industry, energy, agriculture, transport, etc.

As mentioned above, eMBB refers to the extended support of conventional MBB through improved peak/average/cell-edge data rates, capacity and coverage. uRLLC is a requirement for emerging critical applications such as industrial internet, smart grids, infrastructure protection, remote surgery and intelligent transportation systems. Last but certainly not least, mMTC is necessary to support the envisioned 5G scenario with tens of billions of connected IoT devices. It is seen that ultra-low

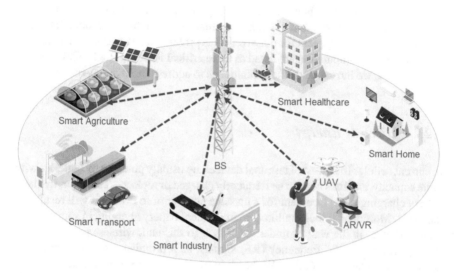

Fig. 1.6 A B5G cellular IoT network

latency, ultra-high efficiency, ultra-high reliability and ultra-high density connectivity are necessary requests of 5G IoT, which in turn accelerate the implementation of key technologies for cellular IoT, e.g., NB-IoT based on the 5G cellular network that has enabled various new applications, such as VR, AR, autonomous driving and etc. However, these applications require ultra-high accuracy of sensing, ultra-low latency of computation, and ultra-high speed of communication among a massive number of IoT devices [23, 24]. Note that although mMTC of 5G cellular IoT emphasizes the number of connections, it dost not demand real-time. Besides, uRLLC of 5G cellular IoT achieves the reliability and timeliness at the cost of low spectrum efficiency and a small number of connections. In this context, it is desired to integrate cellular IoT into B5G wireless networks, namely B5G cellular IoT, c.f. Fig. 1.6, for meeting higher demands with massive access [25].

1.2 Energy, Communication and Computation in B5G Cellular IoT

In general, IoT devices equipped with sensors in B5G cellular IoT collect information from the surroundings or human, then transmit it to the base station (BS) for further decoding, computation, or analysis. The signals from the IoT devices usually have two functions, one for data aggregation based on multiple devices' signals, and the other for information transmission based on individual device's signal. Thus, computation and communication can be abstracted as two elementary tasks of B5G cellular IoT. However, it is not trivial to carry out the two tasks with limited wireless resources. Moreover, due to the high human cost and the environmental strain, frequent battery replacement for massive IoT is prohibitive. Therefore, energy, communication and computation are listed as three critical issues in B5G cellular IoT. In what follows, we introduce the key techniques to address these three crucial factors.

1.2.1   Wireless Energy

In current cellular IoT, most of terminal devices are usually powered by batteries with finite capacity, and thus need to be frequently charged or replaced. However, frequent battery charging or replacement for a massive number of IoT devices will result in a high cost. Moreover, it looks unlikely to implement battery charging in some certain scenarios, e.g., in the wall or under the water. To this end, wireless power transfer (WPT) based on radio frequency (RF) signals, as a promising solution, has been applied to cellular IoT [26, 27]. As shown in Fig. 1.7, the electromagnetic energy of RF signals is harvested and transformed as the electric energy at the IoT devices via WPT, further for prolonging their lifetimes and conducting the sequent tasks. Since WPT can achieve one-to-many charging simultaneously by exploiting the broadcast nature of wireless channels, it is especially suitable for B5G cellular IoT [28, 29]. Due to the signal carrier associated path loss and channel fading, the low efficiency has become a bottleneck of WPT. In this context, energy beamforming was proposed in [30, 31], for improving the efficiency of WPT over fading channels. Hence, WPT becomes a key technique to realize sustainable cellular IoT in B5G wireless networks.

Fig. 1.7 The diagram of WPT

1.2.2 Wireless Communication

The B5G cellular IoT is expected to support seamless access of a massive number of IoT devices with different QoS requirements. However, it is not trivial to achieve massive access over limited radio spectrum. Conventional orthogonal multiple access (OMA) techniques, e.g., frequency division multiple access (FDMA) and time division multiple access (TDMA), allocate a time-frequency resource block to an unique device, resulting in a low spectral efficiency. In other words, OMA techniques can not support massive access with limited wireless resource. In this case, non-orthogonal multiple access (NOMA), as an enabling massive access technique, has received great attention in B5G wireless networks through sharing the time-frequency resource block among multiple devices [32–34], c.f., Fig. 1.8. However, massive NOMA incurs severe co-channel interference, degrading the quality of communication signals [35, 36]. To tackle this issue, spatial beamforming is usually utilized to combat co-channel interference as well as improve the system performance [37, 38]. Especially in B5G cellular IoT, the BS equipped with a large-scale antenna array has ultra-high spatial degrees of freedom to further mitigate co-channel interference [39]. Hence, massive NOMA becomes a key technique to realize reliable communications of IoT devices in B5G cellular IoT.

1.2.3 Wireless Computation

Driven by the huge demands of fast data aggregation for IoT scenarios, B5G cellular IoT is converting from a data-centric network to a computation-centric one. The advanced information processing technologies, such as artificial intelligence (AI) and data mining, will provide ubiquitous computing and intelligent services to effectively realize analysis and processing of massive data from IoT devices [40, 41]. In the future, we may be more concerned about the computation results of the data, such as the sum and the average, rather than the individual data itself. For example,

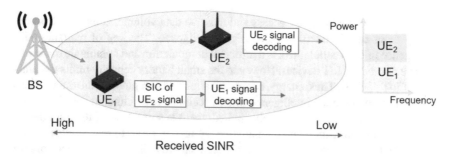

Fig. 1.8 The diagram of NOMA in power domain

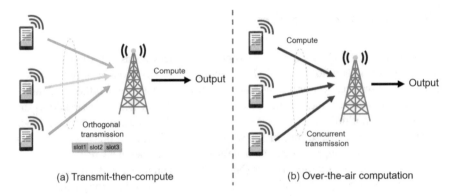

Fig. 1.9 The comparison of conventional computation and AirComp

an IoT-based humidity monitoring system is only interest in the average of humidity in certain region, instead of collecting all observations from sensors. For realizing massive data computation from IoT devices, the conventional approach of transmit-then-compute is no longer applicable for B5G cellular IoT due to the excessively high latency and low spectrum utilization. To address this issue, a promising solution called over-the-air computation (AirComp) has been proposed and caused wide concern [42, 43], which exploits the superposition property of wireless multiple-access channel (MAC) to compute a class of nomographic functions of distributed data from IoT devices via concurrent transmission, c.f. Fig. 1.9. Especially in B5G cellular IoT, AirComp can combine with multiple-input and multiple-output (MIMO) techniques, namely MIMO AirComp, by using spatial degrees of freedom offered by large-scale antenna arrays to spatially multiplex multi-function computation, and can further decrease computation errors by using spatial beamforming [44, 45]. Hence, Air-Comp becomes a key technique to realize efficient computation in B5G cellular IoT.

1.3   Objective of this Book

In the context of massive IoT access and massive data volume, cellular IoT is converting from centric intelligence to edge intelligence. The key of realizing edge intelligence in B5G cellular IoT is to efficient computation and communications of a massive number of IoT devices. However, the small battery capacity limits the functions of IoT devices. Thus, energy, communication and computation are three critical factors of realizing edge intelligence in B5G cellular IoT. Although WPT, NOMA and AirComp have been extensively studied, there is still a lack of the joint design of energy, communication and computation. In fact, in order to realize edge intelligence with limited wireless resources, it is necessary to build a unified framework for energy, communication and computation. To this end, the objective of this book is to provide a comprehensive investigation of the convergence of energy, communication and computation in B5G cellular IoT, and thus provides valuable and useful insights for the analysis, design and optimization of B5G cellular IoT. The main contents of this book are as follows.

In Chap. 2, we study the issue of convergence of energy and communication in B5G cellular IoT with a massive number of IoT devices. In particular, we consider a practical scenario of a sustainable B5G cellular IoT enabled by simultaneous wireless information and power transfer (SWIPT), where the IoT devices have a non-linear energy harvesting (EH) receiver and perform imperfect successive interference cancellation (SIC) due to a limited capability. To realize efficient convergence of energy and communication in B5G cellular IoT, the key is to effectively coordinate the co-channel interference due to non-orthogonal transmission. This is because co-channel interference has two sides of effects on the performance of B5G cellular IoT. On the one hand, co-channel interference decreases the quality of received signals for information decoding (ID). On the other hand, co-channel interference increases the amount of received signals for EH. In general, spatial beamforming and power allocation are utilized to coordinate the co-channel interference. It is well known that the availability of channel state information (CSI) at the BS is the key to perform spatial interference coordination. Without loss of generality, we design the B5G cellular IoT based on three different CSI models including full CSI, imperfect CSI with channel quantization error bounded in an ellipsoid, and imperfect CSI with channel estimation error modeled by Gaussian stochastic process. Corresponding optimization algorithms are proposed to effectively alleviate the impacts of adverse factors as well as improve the overall performance. Extensive simulation results are presented to validate the effectiveness of the proposed algorithms.

In Chap. 3, we focus on the issue of convergence of energy and computation in B5G cellular IoT. Especially, we consider a computation-centric B5G cellular IoT network operated in time division duplex mode (TDD), where a multi-antennas BS plays two roles, i.e., a power beacon in the downlink and a data fusion center in the uplink for multi-modal IoT devices equipped with multiple antennas. At first, the BS utilizes the WPT technique to charge IoT devices via energy beamforming. Then, all IoT devices transmit a set of multi-modal data to the BS simultaneously with the harvested energy. Enabled by the AirComp technique, the BS designs a computation receiver to recover the targeted signal directly. Moreover, in B5G cellular IoT, MIMO AirComp can be utilized to spatially multiplex multi-function computation by exploiting spatial degrees of freedom provided by large-scale antenna arrays, and can further reduce computation errors by using spatial beamforming. Hence, the key to design of B5G cellular IoT lies in beamforming optimization. It is known that beamforming design is closely linked to the CSI. However, in B5G cellular IoT with massive access, it is only able to obtain partial or even no CSI. In other words, it is necessary to take the uncertainty of CSI into consideration for beamforming design, namely robust beamforming. In this context, in order to realize efficient convergence of energy supply and data aggregation in B5G cellular IoT, a robust design algorithm is provided by jointly optimizing beamforming of both WPT and AirComp. Numerical simulation results validate the robustness and effectiveness of the proposed algorithm over the baseline ones.

In Chap. 4, we investigate the issue of convergence of communication and computation in B5G cellular IoT, where each IoT device has two independent signals, one for computation, and the other for communication. For communication, highly accurate sensing information at IoT devices are sent to the BS through by using non-orthogonal communication over wireless multiple access channels. Meanwhile, for computation, AirComp is adopted to substantially reduce latency of massive data aggregation via exploiting the superposition property of wireless multiple-access channels. Specifically, each IoT device carries out beamforming for coordinating the communication signal and the computation signal to be transmitted respectively, and sends a superposition coded signal to the BS over the uplink channel. On the one hand, enabled by the AirComp technique, the BS receives the computation results directly via concurrent data transmission without recovering individual data, and then utilizes a computation receiver to obtain the targeted function signal. On the other hand, the BS decodes the sensing signals of each device through communication receivers. To achieve effective integration of computation and communication under practical but adverse conditions, a robust algorithm is proposed by jointly optimizing transmit power and receive beamforming, with the goal of minimizing the computation error of computation signals while guaranteeing the requirement of communication signals. Extensive simulation results show that the proposed robust algorithm can realize efficient convergence of communication and computation with limited wireless resources.

In Chap. 5, we concentrate on a sustainable B5G cellular IoT integrating energy, communication and computation, where a BS equipped with a large-scale antenna array serves a massive number of multiple-antennas IoT devices. IoT devices equipped with sensors collect information from the surroundings or human, then transmit it to the BS for further decoding, computation, or analysis. Thus, the signals from the IoT devices have two functions, one for data aggregation based on multiple devices' signals, and the other for information transmission based on individual device's signal. However, it is not trivial to carry out the two tasks with limited wireless resources. In this case, NOMA is applied into cellular IoT to realize seamless access of a massive number of devices. Meanwhile, MIMO AirComp is used to reduce latency of massive data aggregation by exploiting the superposition property of wireless multiple-access channel. To effectively address the critical issues of B5G cellular IoT, i.e., energy supply, data aggregation and information transmission, we design a comprehensive framework. Firstly, the BS charges massive IoT devices simultaneously via the WPT technique in the downlink. Secondly, IoT devices with harvested energy carry out the computation task and the communication task in the uplink via AirComp and non-orthogonal transmission over the same spectrum. To improve the overall performance of energy, communication and computation, we propose a joint beamforming design algorithm for the BS and the IoT devices with the goal of minimizing the computation distortion, while guaranteeing the SINR requirements of communication signals. Simulation results validate the effectiveness of the proposed algorithm in B5G cellular IoT.

Finally, in Chap. 6, we give a summary about convergence of energy, communication and computation in B5G cellular IoT. Moreover, we provide several future research directions for realizing edge intelligence in B5G cellular IoT.

References

1. Al-Fuqaha A, Guizani M, Mohammadi M, Aledhari M, Ayyash M (2015) Internet of things: a survey on enabling technologies, protocols, and applications. IEEE Commun Surv Tutor 17(4):2347–2376
2. Xu LD, He W, Li S (2014) Internet of things in industries: a survey. IEEE Trans Ind Inf 10(4):2233–2243
3. Zanella A, Bui N, Castellani A, Vangelista L, Zorzi M (2014) Internet of things for smart cities. IEEE Internet Things J 1(1):22–23
4. Statista Research Department, Internet of Things—number of connected devices worldwide 2015–2025 (2019). https://www.statista.com/statistics/471264/iot-numberof-connected-devices-worldwide
5. Usman R, Kulkarni P, Sooriyabandara M (2017) Low power wide area networks: an overview. IEEE Commun Surv Tutor 19(2):855–873
6. Ikpehai A, Adebisi B, Rabie KM, Anoh K, Ande RE, Hammoudeh M, Gacanin H, Mbanaso UM (2019) Low-power wide area network technologies for internet-of-things: a comparative review. IEEE Commun Surv Tutor 6(2):2225–2240
7. Vejlgaard B, Lauridsen M, Nguyen H, Kovacs IZ, Mogensen P, Sorensen M (2017) Coverage and capacity analysis of Sigfox, LoRa, GPRS, and NB-IoT. In: Proceedings of IEEE VTC spring, Sydney, Australia, pp 4–7
8. Shao X, Chen X, Zhong C, Zhao J, Zhao Z (2019) A unified design of massive access for cellular internet of things. IEEE Internet of Things J 6(2):3934–3947
9. Jia R, Chen X, Zhong C, Ng DWK, Lin H, Zhang Z (2019) Design of non-orthogonal beamspace multiple access for cellular internet-of-things. IEEE J Sel Area Commun 13(3):538–552
10. Chen X (2019) Massive access for cellular internet of things theory and technique. Springer, Singapore
11. Ge X, Qiu Y, Chen J, Huang M, Xu H, Xu J, Zhang W, Yang Y, Wang C-X, Thompson J (2016) Wireless fractal cellular networks. IEEE Wirel Commun 23(5):110–119
12. Sulyman AI, Oteafy SMA, Hassanein HS (2017) Expanding the cellular-IoT umbrella: an architectural approach. IEEE Wirel Commun 24(3):66–71
13. Palattella MR, Dohler M, Grieco A, Rizzo G, Torsner J, Engel T, Ladid L (2016) Internet of things in the 5G era: enablers, architecture, and business models. IEEE J Sel Areas Commun 34(3):510–527
14. Lippuner S, Weber B, Salomon M, Korb M, Huang Q (2018) EC-GSM-IoT network synchronization with support for large frequency offsets. In: Proceeding of IEEE WCNC, Barcelona, Spain, 15–18
15. Soltanmohammadi E, Ghavami K, Naraghi-Pour M (2016) A survey of traffic issues in machine-to-machine communications over LTE. IEEE Internet Things J 3(6):865–884
16. Li Y, Cheng X, Cao Y, Wang D, Yang L (2018) Smart choice for the smart grid: Narrowband internet of things (NB-IoT). IEEE Internet Things J 5(3):1505–1515
17. Ghavimi F, Chen HH (2015) M2M communications in 3GPP LTE/LTE-A networks: architectures, service requirements, challenges, and applications. IEEE Commun Surv Tutor 17(2):525–549
18. Andrews JG, Buzzi S, Choi W, Hanly SV, Lozano A, Soong ACK, Zhang JC (2014) What will 5G be? IEEE J Sel Areas Commun 32(6):1065–1082

19. Wang C-X, Haider F, Gao X, You X-H, Yang Y, Yuan D, Aggoune HM, Haas H, Fletcher S, Hepsaydir E (2014) Cellular architecture and key technologies for 5G wireless communication networks. IEEE Commun Mag 52(2):122–130
20. Agiwal M, Roy A, Saxena N (2016) Next generation 5G wireless networks: a comprehensive survey. IEEE Commun Surv Tutor 18(3):1617–1655
21. Wong VWS, Schober R, Ng DWK, Wang L-C (2017) Key technologies for 5G wireless systems. Cambridge University Press, Cambridge, UK
22. ITU-R, IMT vision—framework and overall objectives of the future development of IMT for 2020 and beyond (2015) Recommendation M.2083-0
23. Wang M, Yang W, Zou J, Ren B, Hua M, Zhang J, You X (2016) Cellular machine-type communications: physical challenges and solutions. IEEE Wirel Commun 23(2):126–135
24. Dawy Z, Saad W, Ghosh A, Andrews JG, Yaacoub E (2017) Toward massive machine type cellular communications. IEEE Wirel Commun 24(1):120–128
25. Chen X, Ng DWK, Yu W, Larsson EG, Al-Dhahir N, Schober R (2020) Massive access for 5G and beyond. IEEE J Sel Area Commun (99):1–24
26. Zhou X, Zhang R, Ho CK (2013) Wireless information and power transfer: architecture design and rate-energy tradeoff. IEEE Trans Commun 61(11):4754–4767
27. Chen X, Yuen C, Zhang Z (2014) Wireless energy and information transfer tradeoff for limited feedback multi-antenna systems with energy beamforming. IEEE Trans Vehic Technol 63(1):407–412
28. Chen X, Wang X, Chen X (2013) Energy-efficient optimization for wireless information and power transfer in large-scale MIMO systems employing energy beamforming. IEEE Wirel Commun Lett 2(6):667–670
29. Chen X, Zhang Z, Chen H-H, Zhang H (2015) Enhancing wireless information and power transfer by exploiting multi-antenna techniques. IEEE Commun Mag 53(4):133–141
30. Na W, Park J, Lee C, Park K, Kim J, Cho S (2018) Energy-efficient mobile charging for wireless power transfer in internet of things networks. IEEE Internet of Things J 5(1):79–92
31. Qi Q, Chen X (2019) Wireless powered massive access for cellular internet of things with imperfect SIC and non-linear EH. IEEE Internet of Things J 6(2):3110–3120
32. Shirvanimoghaddam M, Dohler M, Johnson SJ (2017) Massive non-orthogonal multiple access for cellular IoT: potentials and limitations. IEEE Commun Mag 55(9):55–61
33. Shirvanimogaddam M, Condoluci M, Dohler M, Johnson SJ (2017) On the fundamental limits of random non-orthogonal multiple access in cellular massive IoT. IEEE J Sel Areas Commun 35(10):2238–2252
34. Tian F, Chen X (2019) Multiple-antenna techniques in nonorthogonal multiple access: a review. Front Inform Technol Electron Eng 20(12):1665–1697
35. Chen X, Zhang Z, Zhong C, Jia R, Ng DWK (2018) Fully non-orthogonal communication for massive access. IEEE Trans Commun 16(4):6766–6778
36. Chen X, Jia R (2018) Exploiting rateless coding for massive access. IEEE Trans Vehic Technol 67(11):11253–11257
37. Qi Q, Chen X, Lei L, Zhong C, Zhang Z (2019) Outage-constrained robust design for sustainable B5G cellular Internet of Things. IEEE Trans Wirel Commun 18(12):5780–5790
38. Chen X, Jia R, Ng DWK (2019) On the design of massive non-orthogonal multiple access with imperfect successive interference cancellation. IEEE Trans Commun 67(3):2539–2551
39. Chen X, Zhang Z, Zhong C, Ng DWK (2017) Exploiting multiple-antenna techniques for non-orthogonal multiple access. IEEE J Sel Areas Commun 35(10):2207–2220
40. Li R, Zhao Z, Zhou X, Ding G, Chen Y, Wang Z, Zhang H (2017) Intelligent 5G: When cellular networks meet artificial intelligence. IEEE Wirel Commun 24(5):175–183
41. Zhang H, Li J, Wen B, Xun Y, Liu J (2018) Connecting intelligent things in smart hospitals using NB-IoT. IEEE Internet Things J 5(3):1550–1560
42. Abari O, Rahul H, Katabi D (2016) Over-the-air function computation in sensor networks. http://arxiv.org/pdf/1612.02307.pdf

43. Chen L, Zhao N, Chen Y, Yu FR, Wei G (2018) Over-the-air computation for IoT networks: computing multiple functions with antenna arrays. IEEE Internet of Things J 5(6):5296–5306
44. Li X, Zhu G, Gong Y, Huang K (2019) Wirelessly powered data aggregation for IoT via over-the-air function computation: beamforming and power vontrol. IEEE Trans. Wirel Commun 18(7):3437–3452
45. Qi Q, Chen X, Lei L, Zhong C, Zhang Z (2019) Robust convergence of energy and computation for B5G cellular internet of things. In: Proceeding of IEEE GLOBECOM, Waikoloa, USA, 1–6

Chapter 2
Convergence of Energy and Communication in B5G Cellular Internet of Things

Abstract In this chapter, we investigate the issue of convergence of energy and communication in B5G cellular IoT with a massive number of access wireless devices. Especially, we consider a practical scenario of the B5G cellular IoT, where the IoT devices have a non-linear EH receiver and perform imperfect successive interference cancellation (SIC) due to a limited capability. The benefits offered by a multiple-antenna BS are exploited to enhance the efficiency of both information transmission and power transfer. Considering channel state information (CSI) obtained by the transmitter, we design three different frameworks from the perspectives of full CSI, imperfect CSI with channel quantization error bounded in an ellipsoid, and imperfect CSI with channel estimation error modeled by Gaussian stochastic process, respectively. Corresponding optimization algorithms are proposed to effectively alleviate the impacts of adverse factors as well as improve the overall performance. Extensive simulation results are presented to validate the effectiveness of the proposed algorithms.

2.1 Introduction

Compared to NB-IoT, B5G cellular IoT has the characteristics of massive connectivity and wide coverage [1]. Yet, it is not a trivial task to realize seamless access of a massive number of IoT devices distributed in a wide range with limited wireless resources. On the one hand, traditional orthogonal multiple access schemes, e.g., orthogonal frequency division multiple access (OFDMA), cannot support massive connectivity over limited radio spectrum. To this end, NOMA schemes should be applied in B5G cellular IoT [2, 3]. However, NOMA leads to severe co-channel interference inevitably, which makes it difficult to achieve a wide coverage. To solve this problem, NOMA combining with MIMO techniques is adopted to mitigate the interference by joint spatial beamforming at the BS and SIC at the devices [4, 5]. Especially, since the BS of B5G wireless networks will be equipped with a large-scale antenna array, it is likely to utilize massive MIMO NOMA technique to improve the quality of the received signal significantly [6, 7]. On the other hand, a wide coverage requires a high transmit power, resulting in a high power consumption.

© The Author(s), under exclusive license to Springer Nature Singapore Pte Ltd. 2020
X. Chen and Q. Qi, *Convergence of Energy, Communication and Computation in B5G Cellular Internet of Things*, SpringerBriefs in Electrical and Computer Engineering, https://doi.org/10.1007/978-981-15-4140-7_2

Consequently, the batteries of IoT devices have to be replaced frequently. Nevertheless, battery replacement for a massive number of IoT devices is unaffordable. In this situation, the convergence of energy and communication, namely wireless powered communication, is introduced into IoT [8]. Specifically, the energy signal is broadcasted over the downlink channels, and thus a massive number of IoT devices can be charged simultaneously [9–11]. Therefore, sustainable MIMO NOMA is an enabling solution for achieving the goals of massive connectivity and wide coverage of B5G cellular IoT.

To realize efficient convergence of energy and communication in B5G cellular IoT, the key is to effectively coordinate the co-channel interference due to non-orthogonal transmission [12]. This is because co-channel interference has two sides of effects on the performance of B5G cellular IoT. On the one hand, co-channel interference decreases the quality of received signals for information decoding (ID). On the other hand, co-channel interference increases the amount of received signals for energy harvesting (EH). In general, spatial beamforming and power allocation are utilized to coordinate the co-channel interference [13, 14]. It is well known that to perform spatial interference coordination, the BS has to know channel state information (CSI) [15, 16]. In practice, it is difficult to obtain full CSI of a massive number of IoT devices, especially when the channel dimension is very large due to the deployment of a large-scale antenna array at the BS. In other words, the BS in B5G cellular IoT may be only able to acquire partial CSI because of resource restriction. In [17], a pilot sharing-based CSI acquisition method was designed for massive access systems, and hence the BS was capable of obtaining partial CSI with short pilot sequences. Intuitively, in the case of partial CSI, there exists channel uncertainty at the BS, and then robust schemes have to be designed for guaranteeing quality of services (QoS) of IoT applications [18].

As is well known, there are two robust approaches to overcome the channel uncertainties and satisfy the QoS requirements at each device. The first is a worst-case approach, where the CSI errors are assumed as deterministic vectors within a bounded uncertainty region [19–22], and the second is an outage-constrained approach where CSI errors are modeled as complex Gaussian random vectors [23, 24]. The worst-case approach guarantees the performance with worst CSI errors, and the outage-constrained approach proportionally considers the worst CSI error scenarios by introducing outage probability constraints. In fact, by channel estimation or limited feedback, channel uncertainty at the BS is randomly distributed. Then, the authors in [23] proposed a robust beamforming scheme to minimize the total transmit power based on a stochastic channel uncertainty model in a downlink system. Moreover, a robust secure artificial noise-aided beamforming design method for a cognitive radio system with simultaneous wireless information and power transfer (SWIPT) was analyzed and designed in [24]. However, the both works considered a scenario of a few devices, which might be inapplicable to the B5G cellular IoT with massive access. In this chapter, we consider a general B5G cellular IoT under practical but adverse conditions, where a massive number of simple IoT devices share limited radio spectrum. We provide comprehensive solutions for the convergence of energy and communication in B5G cellular IoT. The contributions of this chapter are three-fold:

1. We consider a practical scenario in B5G cellular IoT, where the simple IoT device has a non-linear EH receiver and carries out imperfect SIC during ID, and give a comprehensive design framework, including signal construction, ID and EH.
2. We analyze the impacts of practically adverse factors, e.g., channel uncertainty due to channel estimation or feedback, imperfect SIC due to low-complexity information receivers, and non-linear EH due to circuit constraints, on the performance of convergence of energy and communication in B5G cellular IoT.
3. We provide three different kinds of design for convergence of energy and communication in B5G cellular IoT, i.e., design with full CSI, worst-case robust design with channel quantization error, and outage-constrained robust design with channel estimation error. Then, corresponding optimization algorithms are proposed to alleviate the impacts of practically adverse factors, and thus effectively improves the overall performance of B5G cellular IoT.

The rest of this chapter is outlined as follows: Sect. 2.2 focuses on the design with full CSI from the perspectives of weighted sum-rate (WSR) maximization and total power consumption (TPC) minimization, respectively. Section 2.3 provides the worst-case robust design algorithms also from the above two aspects. Section 2.4 presents an outage-constrained robust algorithm with the objective of maximizing the WSR. Finally, Sect. 2.5 summarizes the chapter.

Notations: We use bold upper (lower) letters to denote matrices (column vectors), $(\cdot)^H$ to denote conjugate transpose, $\|\cdot\|$ to denote the L_2-norm of a vector, $|\cdot|$ to denote the absolute value, Re{·} to denote the real parts of matrices, E[·] to denote expectation, tr(·) to denote trace of a matrix, Rank(·) to denote rank of a matrix, vec(·) to denote vectorization for a matrix, $\otimes$ to denote the Kronecker product, $\mathbb{C}^{M \times N}$ to denote the set of M-by-N dimensional complex matrix, Pr(·) to denote the probability and $\mathscr{CN}(\mu, \sigma^2)$ to denote the circularly symmetric complex Gaussian (CSCG) distribution with mean μ and variance σ^2.

2.2 Design with Full Channel State Information

In this section, as a warm up, we consider a scenario where the devices in the cellular IoT are fixed, such that CSI keeps unchanged during a relatively long time and the BS is able to obtain full CSI by estimation or feedback at the beginning of each time slot.

2.2.1 System Model

Let us consider a cellular IoT network as shown in Fig. 2.1, where a BS with N_t antennas broadcasts messages to K single antenna IoT UEs. In order to realize efficient massive access over limited radio spectrum, the multiple-antenna NOMA

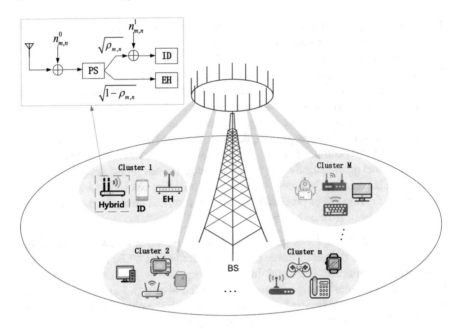

Fig. 2.1 A massive model of the cellular IoT

technique is adopted through combining user clustering and spatial beamforming. In particular, the UEs are partitioned into several clusters in the spatial domain for achieving a balance between system performance and implementation complexity. The UEs in the same cluster share a spatial beam for eliminating the inter-cluster interference, and SIC is carried out in a cluster to reduce the intra-cluster interference.

Without loss of generality, we assume that K UEs are partitioned into M cluster, and the mth cluster contains N_m UEs. The UEs may work in three different modes, namely ID, EH, and hybrid modes. The ID-mode UEs only decode the received signal, while the EH-mode UEs only harvest energy from the received signal. In addition, the hybrid-mode UEs split the received signal into two components, one for ID, and the other for EH. As shown in Fig. 2.1, a PS receiver architecture is deployed at hybrid-mode UEs, and thus the received RF signal at the hybrid-mode UE is divided into two streams by a power splitter in $\rho : 1 - \rho$ proportion, where $0 < \rho < 1$ is the PS ratio. The $1 - \rho$ part of the received RF signal is input into the EH receiver and the rest ρ part is sent to the ID receiver. It is assumed that there are N_m^I ID-mode UEs, N_m^E EH-mode UEs, and N_m^ρ hybrid-mode UEs in the mth cluster, namely $N_m = N_m^\rho + N_m^I + N_m^E$. Since both the ID-mode UEs and the hybrid-mode UEs need to recover the information, we call them as information UEs. Similarly, we call the EH-mode UEs and the hybrid-mode UEs as energy UEs. To simplify expressions, according to the decent order of UEs' channel gain, we denote $\Omega_m^\rho \triangleq \{j_1^\rho, j_2^\rho, \ldots, j_{N_m^\rho}^\rho\}$ as the subscript collection of hybrid-mode UEs, $\Omega_m^I \triangleq \{j_1^I, j_2^I, \ldots, j_{N_m^I}^I\}$ as the subscript collection of information UEs, and $\Omega_m^E \triangleq$

$\{j_1^E, j_2^E, \ldots, j_{N_m^2}^E\}$ as the subscript collection of energy UEs, where $N_m^1 = N_m^\rho + N_m^I$ represents the number of information UEs, and $N_m^2 = N_m^\rho + N_m^E$ is the number of energy UEs.

Based on the available CSI, the BS first performs superposition coding to realize massive spectrum sharing among the K IoT UEs. In general, superposition coding for such a massive NOMA system contains two rounds, i.e., power allocation and transmit beamforming. First, the BS constructs the transmit signal x_m for the mth cluster as below:

$$x_m = \sum_{n \in \Omega_m^I} \sqrt{\alpha_{m,n}} s_{m,n}, \tag{2.1}$$

where $s_{m,n}$ is the Gaussian distributed signal of unit norm for the nth UE in the mth cluster, and $\alpha_{m,n}$ is the intra-cluster power allocation factor with the following constraint:

$$\sum_{n \in \Omega_m^I} \alpha_{m,n} \leq 1, \tag{2.2}$$

Note that intra-cluster power allocation is used to coordinate the intra-cluster interference, and thus improves the performance. Then, the BS constructs the total transmit signal $\mathbf{x}$ as below:

$$\mathbf{x} = \sum_{m=1}^{M} \mathbf{w}_m x_m, \tag{2.3}$$

where $\mathbf{w}_m$ is an N_t-dimensional transmit beam designed for the mth cluster based on CSI. Finally, the BS broadcasts the signal $\mathbf{x}$ over the downlink channels. Therefore, the received signal at the nth UE in the mth cluster can be expressed as

$$y_{m,n} = \mathbf{h}_{m,n}^H \mathbf{x} + n_{m,n}^0, \tag{2.4}$$

where $\mathbf{h}_{m,n}$ denotes the N_t-dimensional channel vector from the BS to the nth UE in the mth cluster and $n_{m,n}^0$ is additive white Gaussian noise (AWGN) with variance σ_0^2. For ID-mode and EH-mode UEs, the received signal $y_{m,n}$ is used for ID and EH directly. For the hybrid-mode UEs, PS is carried out on the received signals. Hence, the signals used for ID and EH are given by $\sqrt{\rho_{m,n}} y_{m,n}$ and $\sqrt{1 - \rho_{m,n}} y_{m,n}$, respectively.

At the information UEs, including ID-mode UEs and hybrid-mode UEs, SIC is performed to decrease the intra-cluster interference caused by non-orthogonal transmission. In practice, each UE knows its effective channel gain through channel estimation for coherent detection [25, 26]. Thus, the UEs can convey the effective channel gains to the BS via the uplink channel and the BS determines the order of effective channel gains in each cluster which is informed to the UEs through the downlink channel. It is assumed that the effective channel gains in the mth cluster have the following order:

$$|\mathbf{h}_{m,j_1^I}^H \mathbf{w}_m|^2 \geq \cdots \geq |\mathbf{h}_{m,j_{N_m^1}^I}^H \mathbf{w}_m|^2, \tag{2.5}$$

Then, according to the principle of SIC, the j_i^Ith information UE first decodes and cancels the interfering signals related to the $j_{N_m^I}^I$th to j_{i+1}^Ith information UEs. If SIC is performed perfectly, the intra-cluster interference from the UEs with weaker channel gains can be cancelled completely. However, in practical systems, due to the hardware limitation of the IoT UEs, the low signal quality and other factors, the decoding error of the weak interference signal might occur. Thus, there still exists residual interference from the weak UEs even after SIC. In general, the residual interference caused by imperfect SIC is a complicated function of multiple factors, e.g., coding/modulation related parameters, channel related issues (fading and shadowing), device/hardware/battery related restrictions, etc. Moreover, due to the characteristics of error propagation of imperfect SIC, it is difficult to model the impact of imperfect SIC. In [27], a linear model is given to show the impact of imperfect SIC. Under such a model for imperfect SIC, the post-SIC signal at the nth information UE in the mth cluster for ID can be given by

$$y_{m,n}^{ID} = \sqrt{\psi_{m,n}}(\mathbf{h}_{m,n}^H \mathbf{x} + n_{m,n}^0) + \phi_{m,n} n_{m,n}^1$$

$$= \sqrt{\psi_{m,n}} \mathbf{h}_{m,n}^H \mathbf{w}_m \sqrt{\alpha_{m,n}} s_{m,n} + \sqrt{\psi_{m,n}} \mathbf{h}_{m,n}^H \mathbf{w}_m \sum_{i=j_1^I}^{j_{n-1}^I} \sqrt{\alpha_{m,i}} s_{m,i} \tag{2.6}$$

$$+ \sqrt{\psi_{m,n} \eta_{m,n}} \mathbf{h}_{m,n}^H \mathbf{w}_m \sum_{i=j_{n+1}^I}^{j_{N_m^1}^I} \sqrt{\alpha_{m,i}} s_{m,i} \sqrt{\psi_{m,n}} \mathbf{h}_{m,n}^H \sum_{j=1,j\neq m}^{M} \mathbf{w}_j \sum_{i\in\Omega_j^I} \sqrt{\alpha_{j,i}} s_{j,i}$$

$$+ \sqrt{\psi_{m,n}} n_{m,n}^0 + \phi_{m,n} n_{m,n}^1, \tag{2.7}$$

where $\psi_{m,n}$ and $\phi_{m,n}$ are used to distinguish UEs' modes, which are defined as follows:

$$\begin{cases} \psi_{m,n} = 1 \text{ and } \phi_{m,n} = 0, & \text{ID-mode UE} \\ \psi_{m,n} = \rho_{m,n} \text{ and } \phi_{m,n} = 1, & \text{Hybrid-mode UE} \end{cases} \tag{2.8}$$

Moreover, the variable $\eta_{m,n}$ denotes the coefficient of imperfect SIC at the nth information UE in the mth cluster, which can be obtained by long-term measurement.[1] $n_{m,n}^1 \sim \mathcal{CN}(0, \sigma_1^2)$ denotes the baseband AWGN caused by the RF band to baseband signal conversion, and σ_1^2 denotes the noise power. As a result, the received signal-to-interference-plus-noise ratio (SINR) at the ID receiver of the information UEs can be given by

---

[1] By off-line measuring a large number of samples in a long training time, the residual interference can be accurately approximated using a Gaussian distribution due to the central limit theorem, and the variance is a function of the received power [43]. Then, by comparing the powers of the residual interference and the received signal, the coefficient $\eta_{m,n}$ can be obtained at the BS.

$$\Gamma_{m,n} = \frac{\left|\mathbf{h}_{m,n}^H \mathbf{w}_m\right|^2 \alpha_{m,n}}{\sum\limits_{j=1}^{M} \sum\limits_{i \in \Omega_m^I} \kappa_{j,i}^{m,n} \alpha_{j,i} \left|\mathbf{h}_{m,n}^H \mathbf{w}_j\right|^2 + \sigma_0^2 + \frac{\phi_{m,n}}{\psi_{m,n}} \sigma_1^2}, \tag{2.9}$$

where $\kappa_{j,i}^{m,n}$ just is an auxiliary parameter, which is given by

$$\kappa_{j,i}^{m,n} = \begin{cases} \eta_{m,n}, & \text{if } j = m \text{ and } i > n, \\ 0, & \text{if } j = m \text{ and } i = n, \\ 1, & \text{otherwise.} \end{cases} \tag{2.10}$$

Note that the auxiliary parameter $\kappa_{j,i}^{m,n}$ represents the impact of imperfect SIC on the design of the wireless powered cellular IoT network. First, we use the parameter $\kappa_{j,i}^{m,n} = 1, \forall j = m$ and $i > n$ to represent the situation where some certain IoT devices cannot perform SIC. Second, we represent the ideal situation under perfect SIC by letting $\kappa_{j,i}^{m,n} = 0, \forall j = m$ and $i > n$. Finally, if the IoT devices perform SIC but SIC is imperfect, the parameter $\kappa_{j,i}^{m,n}$ is within the range $(0, 1)$. Thus, the adopted model is applicable in various cases of SIC.

Then, the received RF signal at the EH receiver of the energy UEs, including EH-mode UEs and hybrid-mode UEs, can be expressed as

$$y_{m,n}^{\text{EH}} = \sqrt{1 - \psi_{m,n}} (\mathbf{h}_{m,n}^H \mathbf{x} + n_{m,n}^0), n \in \Omega_m^E \tag{2.11}$$

where

$$\begin{cases} \psi_{m,n} = 0, & \text{EH-mode UE} \\ \psi_{m,n} = \rho_{m,n}, & \text{Hybrid-mode UE} \end{cases} \tag{2.12}$$

Thus, the input RF power for the EH receiver at the energy UEs is given by

$$P_{m,n}^{\text{in}} = (1 - \psi_{m,n}) \sum_{j=1}^{M} \left|\mathbf{h}_{m,n}^H \mathbf{w}_j\right|^2, \tag{2.13}$$

If the traditional linear EH model is adopted [28, 29], the harvested energy at the energy UEs can be expressed as

$$P_{m,n}^{(L)} = \varepsilon_{m,n} P_{m,n}^{\text{in}} = \varepsilon_{m,n} (1 - \psi_{m,n}) \sum_{j=1}^{M} \left|\mathbf{h}_{m,n}^H \mathbf{w}_j\right|^2, \tag{2.14}$$

where $\varepsilon_{m,n} \in (0, 1]$ describes the energy conversion efficiency. In general, for the conventional linear EH model, the energy conversion efficiency is independent of the input power level at the energy receiver (ER). However, the EH circuit presents nonlinear characteristics in practice. In other words, the total harvested energy at the ER is not linearly proportional to the received RF power. Specifically, the RF energy conversion efficiency improves as the input power increases in the low power

Fig. 2.2 A comparison of
the harvested power between
the linear EH model (2.14)
and non-linear EH model
(2.15)

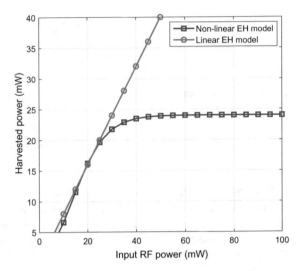

region, but the harvested energy will be saturated if the power is high enough. It is intuitive that the conventional linear EH model is not accurate especially when the received power is high, which may cause serious performance loss. Thus, in order to accurately analyze and optimize performance, we adopt a non-linear EH model to obtain the dynamic RF energy conversion efficiency of practical EH circuit according to a previous related work [30], which captures the effect of the non-linear phenomena caused by hardware constraints. Base on this nonlinear EH model, the harvested energy at the energy UEs is given by

$$
P_{m,n}^{(NL)} = \frac{\frac{M_{m,n}}{1+\exp(-a_{m,n}(P_{m,n}^{in}-b_{m,n}))} - \frac{M_{m,n}}{1+\exp(a_{m,n}b_{m,n})}}{1 - \frac{1}{1+\exp(a_{m,n}b_{m,n})}},
\tag{2.15}
$$

where $M_{m,n}$ is a constant denoting the maximum harvested power at the nth energy UE in the mth cluster when the EH circuit is saturated. Additionally, parameters $a_{m,n}$ and $b_{m,n}$ are constants related to the detailed circuit specifications, which account for physical hardware phenomena, such as circuit sensitivity limitations and leakage currents [31, 32]. Figure 2.2 illustrates the harvested power of linear EH model and non-linear EH model under the same conditions, where we set $a_{m,n} = 150$, $b_{m,n} = 0.014$, $M_{m,n} = 24\,\text{mW}$, and $\varepsilon_{m,n} = 0.8$. It can be seen that the harvested power of the non-linear EH model will gradually approach saturated with the increment of the input RF power, while the harvested energy of the linear one continues to grow linearly. Thus, the non-linear model significantly affects the harvested energy, especially in the region of high input RF power.

As seen in (2.9) and (2.15), spatial beam, transmit power, and PS ratio have great effects on the performance of wireless powered cellular IoT. Thus, it is desired to optimize these system parameters to improve the overall performance. Especially in the practical scenario, it makes sense to alleviate the impacts of imperfect SIC and non-linear EH through performance optimization.

2.2.2 Problem Formulation and Optimization Design

In this section, we aim to jointly design spatial beam, transmit power, and PS ratio for the considered wireless powered cellular IoT network with the non-linear EH model in the presence of imperfect SIC. As is well known that the maximization of WSR and the minimization of the TPC are two commonly used design objectives in wireless communications. Therefore, we also design the wireless powered cellular IoT network based on the two objectives respectively.

2.2.2.1 Weighted Sum Rate Maximization Design

The design with the goal of maximizing the WSR of the information UEs in the wireless powered cellular IoT network while guaranteeing the energy harvesting requirements of the energy UEs can be formulated as the following optimization problem:

$$\text{OP1} : \max_{\mathbf{w}, \boldsymbol{\rho}, \boldsymbol{\alpha}} \sum_{m=1}^{M} \sum_{n \in \Omega_m^I} \theta_{m,n} R_{m,n}$$

$$\text{s.t. } \text{C1} : \sum_{m=1}^{M} \|\mathbf{w}_m\|^2 \leq P_{\max},$$

$$\text{C2} : \sum_{n \in \Omega_m^I} \alpha_{m,n} \leq 1, \forall m,$$

$$\text{C3} : \alpha_{m,n} \leq \alpha_{m,n+1}, n \in \Omega_m^I,$$

$$\text{C4} : P_{m,n}^{(\text{NL})} \geq q_{m,n}, \forall m, n \in \Omega_m^E,$$

$$\text{C5} : 0 < \rho_{m,n} < 1, \forall m, n \in \Omega_m^\rho$$

where $R_{m,n} = \log_2(1 + \Gamma_{m,n})$ is the achievable rate (in b/s) of the nth information UE in the mth cluster, $\theta_{m,n} > 0$ denotes the priority, $P_{\max} > 0$ is the maximum transmit power budget at the BS, and $q_{m,n} > 0$ is the required minimum amount of harvested energy, respectively. $\mathbf{w} = \{\mathbf{w}_1, \ldots, \mathbf{w}_M\}$, $\boldsymbol{\alpha} = \{\alpha_{1,j_1^I}, \ldots, \alpha_{M,j_{N_m^I}^I}\}$ and $\boldsymbol{\rho} = \{\rho_{1,j_1^\rho}, \ldots, \rho_{M,j_{N_m^\rho}^\rho}\}$ are the collection of spatial beams, power allocation factors and PS ratios, respectively. Note that C1 represents the power allocation between the clusters, C2 denotes the power allocation for the information UEs within a cluster, C3 is used to facilitate SIC in a cluster, C4 is the energy constraint of the energy UEs, and C5 is the constraint condition on the PS ratios. It is clear that OP1 is not a convex problem, and thus it is difficult to obtain the optimal solution directly. To solve this problem, we shall first deal with the nonconvex constraint C4 caused by non-linear EH. Define $Q_{m,n} \triangleq b_{m,n} - \frac{1}{a_{m,n}} \ln \frac{e^{a_{m,n}b_{m,n}}(M_{m,n}-q_{m,n})}{q_{m,n}e^{a_{m,n}b_{m,n}}+M_{m,n}}$, then C4 can be rewritten as

$$C4' : \sum_{j=1}^{M} |\mathbf{h}_{m,n}^{H}\mathbf{w}_{j}|^{2} \geq \frac{Q_{m,n}}{1-\psi_{m,n}}, \tag{2.16}$$

As seen that, the left side of (2.87) is a quadratic-over-linear function, and thus the constraint C4' is still nonconvex. To deal with it, we need to reformulate (2.87) into a convex constraint. Define $\Delta \mathbf{w}_{m} = \mathbf{w}_{m} - \widetilde{\mathbf{w}}_{m}$, then $|\mathbf{h}_{m,n}^{H}\mathbf{w}_{m}|^{2}$ can be expressed as

$$\begin{aligned}
\mathbf{h}_{m,n}^{H}\mathbf{w}_{m}\mathbf{w}_{m}^{H}\mathbf{h}_{m,n} = \mathbf{h}_{m,n}^{H}\widetilde{\mathbf{w}}_{m}\widetilde{\mathbf{w}}_{m}^{H}\mathbf{h}_{m,n} &+\mathbf{h}_{m,n}^{H}\Delta\mathbf{w}_{m}\widetilde{\mathbf{w}}_{m}^{H}\mathbf{h}_{m,n} \\
&+ \mathbf{h}_{m,n}^{H}\widetilde{\mathbf{w}}_{m}\Delta\mathbf{w}_{m}^{H}\mathbf{h}_{m,n} \\
&+ \mathbf{h}_{m,n}^{H}\Delta\mathbf{w}_{m}\Delta\mathbf{w}_{m}^{H}\mathbf{h}_{m,n},
\end{aligned} \tag{2.17}$$

where $\widetilde{\mathbf{w}}_{m}$ represents the last iteration of $\mathbf{w}_{m}$. Ignoring the quadratic term $\mathbf{h}_{m,n}^{H} \Delta\mathbf{w}_{m}\Delta\mathbf{w}_{m}^{H}\mathbf{h}_{m,n}$ in the (2.17), we have

$$\begin{aligned}
\mathbf{h}_{m,n}^{H}\mathbf{w}_{m}\mathbf{w}_{m}^{H}\mathbf{h}_{m,n} \approx \mathbf{h}_{m,n}^{H}\widetilde{\mathbf{w}}_{m}\widetilde{\mathbf{w}}_{m}^{H}\mathbf{h}_{m,n} &+\mathbf{h}_{m,n}^{H}\Delta\mathbf{w}_{m}\widetilde{\mathbf{w}}_{m}^{H}\mathbf{h}_{m,n} \\
&+ \mathbf{h}_{m,n}^{H}\widetilde{\mathbf{w}}_{m}\Delta\mathbf{w}_{m}^{H}\mathbf{h}_{m,n}
\end{aligned} \tag{2.18}$$

Based on (2.18), for a given point $\widetilde{\mathbf{w}}_{j}$, C4' can be approximated as C4'':

$$\sum_{j=1}^{M} \mathbf{h}_{m,n}^{H} \left(\widetilde{\mathbf{w}}_{j}\widetilde{\mathbf{w}}_{j}^{H} + \Delta\mathbf{w}_{j}\widetilde{\mathbf{w}}_{j}^{H} + \widetilde{\mathbf{w}}_{j}\Delta\mathbf{w}_{j}^{H}\right) \mathbf{h}_{m,n} \geq \frac{Q_{m,n}}{1-\psi_{m,n}}, \tag{2.19}$$

It is obvious that C4'' is convex due to its linearity. Next, we deal with the convexity of the objective function of OP1. It is seen that the objective function is a WSR, which is in general non-convex. In order to solve this problem, we transform the objective function according to the following lemma.

Lemma 2.1 *The received SINR $\Gamma_{m,n}$ and the minimum mean squared error (MMSE) $e_{m,n}$ between the transmit and receive signals have the following equivalent relationship:*

$$1 + \Gamma_{m,n} = e_{m,n}^{-1}, \tag{2.20}$$

Proof Please refer to Appendix A.   □

Thus, the objective function of OP1 is reduced as

$$\min_{\mathbf{w},\alpha,\rho} \sum_{m=1}^{M} \sum_{n\in\Omega_{m}^{I}} \theta_{m,n}\log_{2}(e_{m,n}), \tag{2.21}$$

However, the objective function (2.21) is still not convex. Note that (2.21) aims to minimize a function of MMSE, which is equivalent to minimizing a function of MSE for a given MMSE receiver. In other words, the optimization objective in (2.21) can be transformed as

$$\min_{\mathbf{w},\mathbf{v},\alpha,\rho,\beta} \sum_{m=1}^{M} \sum_{n\in\Omega_m^l} \theta_{m,n} \left(\beta_{m,n}\mathrm{MSE}_{\mathrm{m,n}} - \log_2(\beta_{m,n})\right), \tag{2.22}$$

where $\mathbf{v} = \{v_{1,j_1^l}, \ldots, v_{M,j_{N_m^1}^l}\}$ is a collection of the receivers, $\mathrm{MSE}_{m,n}$ is the MSE related to the nth information UE in the mth cluster and $\beta = \{\beta_{1,j_1^l}, \ldots, \beta_{M,j_{N_m^1}^l}\}$ is a collection of auxiliary variables, which makes us further solve this problem [33]. Note that the objective function (2.22) obtains the minimum value only when $\beta_{m,n} = \mathrm{MSE}_{m,n}^{-1}$. Hence, the optimization problem OP1 is changed as

$$\mathrm{OP2}: \min_{\mathbf{w},\mathbf{v},\alpha,\rho,\beta} \sum_{m=1}^{M} \sum_{n\in\Omega_m^l} \theta_{m,n} \left(\beta_{m,n}\mathrm{MSE}_{\mathrm{m,n}} - \log_2(\beta_{m,n})\right),$$

s.t. C1, C2, C3, C4″, C5,

According to the definition of $\mathrm{MSE}_{m,n}$ in (2.98), it is known that OP2 is not a joint convex function of $\mathbf{w}$, $\mathbf{v}$, α, ρ, β, but it is a convex function of each optimization variable. Therefore, we can apply the coordinate gradient decent method to solve OP2 [35]. Specifically, we optimize one variable by fixing the others. The five variables are iteratively optimized until they approach a stationary point. First, the auxiliary variable $\beta_{m,n}$ is always equal to $\mathrm{MSE}_{m,n}^{-1}$ in the sense of maximizing the WSR. Second, for the MMSE receiver, we have a closed-form solution $v_{m,n} = \sqrt{\psi_{m,n}\alpha_{m,n}}\mathbf{w}_m^H\mathbf{h}_{m,n}\mathbf{U}_{m,n}^{-1}$. Since transmit beam $\mathbf{w}_m$, power allocation factor $\alpha_{m,n}$ and PS ratio $\rho_{m,n}$ have multiple linear constraint conditions, they can be iteratively optimized by CVX [36]. In summary, the design for maximizing the WSR of the wireless powered cellular IoT network can be described as Algorithm 2.1.

Remark 2.1 For the approximation (2.17)–(2.19), we denote $\Delta\mathbf{w}_m = \mathbf{w}_m - \tilde{\mathbf{w}}_m$ and utilize a sequential convex approximation (SCA) method to transform the non-convex EH harvested constraint into a linear one, and ignore the quadratic term $\mathbf{h}_{m,n}^H\Delta\mathbf{w}_m\Delta\mathbf{w}_m^H\mathbf{h}_{m,n}$. Since the second-order term is quite small compared to other terms, this approximation is accurate as long as we set the feasible initial value in the first iteration. In the proposed Algorithm 2.1, we initialize $\mathbf{w}_m = \tilde{\mathbf{w}}_m = \sqrt{\frac{P_{\max}}{M}}[1, 0, \cdots, 0]^T$ and make $\tilde{\mathbf{w}}_m = \mathbf{w}_m^*$ after each iteration, so as to ensure that $||\Delta\mathbf{w}_m||$ is small enough.

Algorithm 2.1 Design with Full CSI for WSR Maximization

Input: $a_{m,n}, b_{m,n}, M_{m,n}, q_{m,n}, \eta_{m,n}, \theta_{m,n}, \sigma_0^2, \sigma_1^2, P_{\max}$.
Output: $\mathbf{w}_m, \alpha_{m,n}, \rho_{m,n}$

1: **Initialize** $\mathbf{w}_m = \tilde{\mathbf{w}}_m = \sqrt{\frac{P_{\max}}{M}}[1, 0, \cdots, 0]^T$, $\alpha_{m,n} = \frac{1}{N_m^I}, \forall m, n \in \Omega_m^I, \rho_{m,n} = 0.5, \forall m, n \in$
   Ω_m^ρ, the WSR $R = 0$, convergence accuracy $\Delta = 1$, the maximum number of iterations
   $T_{\max} = 30$, and iteration index $t = 1$.
2: **Set** the auxiliary variable $\beta_{m,n} = \text{MSE}_{m,n}^{-1}$, and the MMSE receiver $v_{m,n} =$
   $\sqrt{\psi_{m,n}\alpha_{m,n}}\mathbf{w}_m^H\mathbf{h}_{m,n}\mathbf{U}_{m,n}^{-1}$
3: **while** $\Delta > 0.01$ and $t < T_{\max}$ **do**
4:    Solve OP2 by CVX with fixed $\rho_{m,n}^{(t)}$ and $\alpha_{m,n}^{(t)}$, then obtain $\mathbf{w}_m^{(t)}$ and update $\tilde{\mathbf{w}}_m^{(t)} = \mathbf{w}_m^{(t)}$;
5:    Solve OP2 by CVX with fixed $\mathbf{w}_m^{(t)}$ and $\rho_{m,n}^{(t)}$, then obtain $\alpha_{m,n}^{(t)}$;
6:    Solve OP2 by CVX with fixed $\mathbf{w}_m^{(t)}$ and $\alpha_{m,n}^{(t)}$, then obtain $\rho_{m,n}^{(t)}$;
7:    Update $\beta_{m,n}^{(t)}, v_{m,n}^{(t)}$ and compute $R^{(t)}$;
8:    Update $\Delta = R^{(t)} - R^{(t-1)}$;
9:    Update $t = t + 1$;
10: **end while**

Complexity Analysis: For simplicity of complexity analysis, let K be the total number of UEs and N_t denote the number of antennas at the BS. The proposed Algorithm 2.1 has the same complexity order as the conventional weighted minimum mean square error (WMMSE) algorithm since it only introduces two additional steps in each iteration, i.e., updating $\beta_{m,n}$ and $v_{m,n}$, which are both closed-form functions of the spatial beam. It is obvious that the main computational complexity of the proposed Algorithm 2.1 comes from obtaining optimal solution in steps 4, 5, and 6. Specifically, problem OP2 is a quadratically constrained quadratic programming (QCQP) problem, yet it can also be equivalently reformulated as a second order cone programming (SOCP) problem. Thus, the complexity of Algorithm 2.1 is equivalent to that of solving a SOCP problem by using the interior point method, which can be approximated as $\mathcal{O}((KN_t)^3)$ [34]. In addition, we list the WSR and the computing time of the proposed Algorithm 2.1 with respect to different SNR values in the Table 2.1, where $N_t = 64$, $K = 48$, $N_m = N = 4$, $N_m^\rho = 2$, $N_m^I = N_m^E = 1$, $\eta = 0.05$, and the convergence criteria $\Delta = 0.01$. Therefore, the proposed Algorithm 2.1 converges to a locally optimal solution in polynomial time [35].

Table 2.1 The computing time at various SNR of Algorithm 2.1

SNR (dB)	0	5	10	15	20	25	30
Sum rate (bit/s/Hz)	19.6532	33.4155	46.6364	56.3802	62.5175	66.8673	70.6953
Time (s)	445.7226	529.0640	698.8932	620.3353	706.2412	627.1261	632.6618

2.2.2.2 Total Power Consumption Minimization Design

In this section, we design the wireless powered cellular IoT network from the perspective of minimizing the TPC at the BS, while meeting QoS requirements. As shown in (2.13), the TPC at the BS is mainly determined by transmit beams. In order to simplify the computational complexity, we design a scheme with fixed-proportion power allocation within each cluster. Following previous related work on power allocation [3], we set $\alpha_{m,n} = \frac{n}{\sum_{i \in \Omega_m^I} i}$, $\forall m, n \in \Omega_m^I$ fixedly for achieving a balance between system performance and SIC implementation.

Although the cellular IoT may have multiple QoS requirements, e.g., the delay, throughput and reliability, they can be represented as a function of SINR. Thus, we consider a minimum SINR requirement here. Meanwhile, due to the demand of wireless charging for IoT UEs, we also consider a minimum energy harvesting requirement. Let $\gamma_{m,n} > 0$ and $q_{m,n} > 0$ be the minimal required SINR threshold and the required minimum amount of harvested energy, respectively. The optimization problem related to the minimization of the TPC can be mathematically expressed as

$$\text{OP3} : \min_{\mathbf{w}, \rho} \sum_{m=1}^{M} \|\mathbf{w}_m\|^2,$$

$$\text{s.t. } \text{C4', C5, C6} : \Gamma_{m,n} \geq \gamma_{m,n}, \forall m, n \in \Omega_m^I$$

In general, OP3 is not convex due to the coupling of optimization variables in C6. To solve this problem, we reformulate OP3 as the following semi-definite programming (SDP) problem:

$$\text{OP4} : \min_{\mathbf{w}, \rho} \sum_{m=1}^{M} \text{tr}(\mathbf{W}_m)$$

$$\text{s.t. } \text{C5, C7} : \text{tr}(\mathbf{H}_{m,n}\mathbf{W}_m)\left(\frac{\alpha_{m,n}}{\gamma_{m,n}} - \sum_{i \in \Omega_m^I} \kappa_{m,i}^{m,n}\alpha_{m,i}\right) - \sum_{j=1, j \neq m}^{M} \text{tr}(\mathbf{H}_{m,n}\mathbf{W}_j) \geq \sigma_0^2 + \frac{\phi_{m,n}\sigma_1^2}{\psi_{m,n}}, n \in \Omega_m^I,$$

$$\text{C8} : \sum_{j=1}^{M} \text{tr}\left(\mathbf{H}_{m,n}\mathbf{W}_j\right) \geq \frac{Q_{m,n}}{1 - \psi_{m,n}}, n \in \Omega_m^E,$$

$$\text{C9} : \text{Rank}(\mathbf{W}_m) = 1, \forall m,$$

$$\text{C10} : \mathbf{W}_m \succeq 0, \forall m,$$

where $\mathbf{H}_{m,n} = \mathbf{h}_{m,n}\mathbf{h}_{m,n}^H$, and $\mathbf{W}_m = \mathbf{w}_m\mathbf{w}_m^H$. Due to the rank-one constraint in C9, OP4 is still non-convex. To this end, by dropping the rank-one matrix constraint C9, we relax OP4 into OP5, which is expressed as

$$\text{OP5} : \min_{\mathbf{w}, \rho} \sum_{m=1}^{M} \text{tr}(\mathbf{W}_m)$$

$$\text{s.t. } \text{C5, C7, C8, C10,}$$

where constraint C5 and C10 are convex sets due to their linearity. In addition, since the Hessian matrices of C7 and C8 satisfy that $\nabla^2_{c7} \succeq 0$ and $\nabla^2_{c8} \succeq 0$ as shown in (2.99), constraint C7 and C8 are also convex sets.

$$\nabla^2_{C7} = \begin{bmatrix} \mathbf{0} & 0 \\ 0 & -\frac{2\phi_{m,n}\sigma_1^2}{\rho_{m,n}^3} \end{bmatrix} \succeq \mathbf{0}, \ \nabla^2_{C8} = \begin{bmatrix} \mathbf{0} & 0 \\ 0 & -\frac{2Q_{m,n}}{(1-\rho_{m,n})^3} \end{bmatrix} \succeq \mathbf{0}. \tag{2.23}$$

Thus, the problem OP5 is a convex problem, which can be easily solved by some off-the-shelf optimization softwares, i.e, CVX. For the solution to OP5, we have the following proposition:

Proposition 2.1 *The optimal solution* $\mathbf{W}_m^*$ *of the problem OP5 satisfies* Rank $(\mathbf{W}_m^*) = 1, \forall m.$

Proof Please refer to Appendix B.

According to Proposition 2.1, it is known that OP4 and OP5 is equivalent. Thus, we can get the unique solution $\mathbf{w}_m^*$ to OP4 by eigenvalue decomposition (EVD) on $\mathbf{W}_m^*$. In summary, the design of the wireless powered cellular IoT network for minimizing the TPC can be described as Algorithm 2.2.

Algorithm 2.2 Design with Full CSI for TPC Minimization

Input: $a_{m,n}, b_{m,n}, M_{m,n}, \eta_{m,n}, \sigma_0^2, \sigma_1^2, q_{m,n}, \gamma_{m,n}, P_{\max}$
Output: $\mathbf{w}_m, \rho_{m,n}$

1: **Initialize** $\mathbf{w}_m = \sqrt{\frac{P_{\max}}{M}}[1, 0, \cdots, 0]^T$ and $\rho_{m,n} = 0.5, \forall m, n \in \Omega_m^\rho$.
2: **Set** $\alpha_{m,n} = \frac{n}{\sum_{i \in \Omega_m^I} i}, \forall m, n \in \Omega_m^I$, and $\mathbf{W}_m = \mathbf{w}_m \mathbf{w}_m^H$.
3: Solve OP5 with CVX, then obtain $\mathbf{W}_m^*$ and $\rho_{m,n}^*$.
4: Based on $\mathbf{W}_m^*$, compute $\mathbf{w}_m^*$ by EVD.

2.2.3 Numerical Results

This section provides some numerical results to validate the effectiveness of the proposed algorithms for the wireless powered cellular IoT network with massive connections. Unless otherwise stated, we set $N_t = 64$, $K = 48$, $M = 12$, $N_m = N = 4$, $N_m^\rho = 2$, $N_m^I = N_m^E = 1$, $\eta_{m,n} = \eta = 0.05$, $\sigma_0^2 = 0.1$, $\sigma_1^2 = 1$ and $\theta_{m,n} = 1$, $\forall m, n$. In addition, we use SNR (in dB) to denote the ratio of transmit power at the BS and the noise variance. For the non-linear EH model, we set $M_{m,n} = 24$ mW, $a_{m,n} = 150$ and $b_{m,n} = 0.014$ according to the practical circuit parameters provided by [30]. For the linear EH model, we set $\varepsilon_{m,n} = \varepsilon = 0.5$. Without loss of generality, all IoT UEs have the same required SINR threshold γ_0 and the same required EH threshold q_0.

Fig. 2.3 Convergence of the
proposed Algorithm 2.1

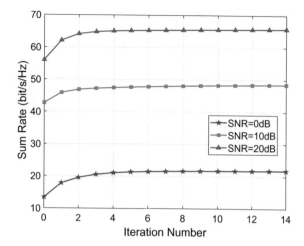

First, we show the convergence behavior of the proposed Algorithm 2.1 with different SNR values in Fig. 2.3. It is seen that the proposed Algorithm 2.1 converges after a few number of iterations under different SNR values. Thus, the proposed Algorithm 2.1 has a fast convergence behavior.

Then, we show the performance gain of the proposed Algorithm 2.1 over several baseline algorithms from the perspective of maximize the sum rate. Specifically, we compare the sum rate of the proposed Algorithm 2.1, a TDMA algorithm, a fixed PS ratio algorithm ($\rho_{m,n} = \rho = 0.5$, $\forall m, n \in \Omega_m^\rho$) and a zero-forcing beamforming (ZFBF) algorithm. From Fig. 2.4, it is seen that the proposed Algorithm 2.1 performs much better than other three baseline algorithms in low and median SNR region, while the ZFBF algorithm performs best in high SNR region. This is because the cellular

Fig. 2.4 Performance
comparison of different
massive access algorithm

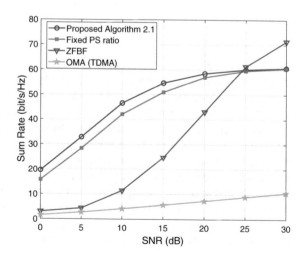

Fig. 2.5 Performance
comparison of different PS
ratios

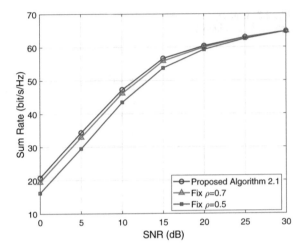

Fig. 2.6 Influence of the
number of UEs in a cluster
on the sum rate

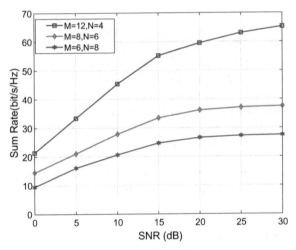

IoT network with massive connections is noise limited when SNR is low, while it is
interference limited as SNR increases. ZFBF algorithm is able to cancel the inter-
cluster interference completely, and thus it can increase the sum rate at high SNR.
Moreover, it is found that the performance gap of the proposed Algorithm 2.1 and
the fixed PS ratio algorithm is small, especially in high SNR region. This is because
the power consumption of the harvested energy is small and the impact of the PS
ratio can be ignored when the transmit power is high. In this context, we continue
to investigate the impact of the coefficient of a fixed PS ratio algorithm as shown
in Fig. 2.5. It is seen that the higher the coefficient ρ, the better performance which
is closer to the proposed Algorithm 2.1. Therefore, for reducing the computational
complexity, it is possible to replace the proposed Algorithm 2.1 with the fixed PS
ratio algorithm by utilizing a large ρ when transmit power is high.

Fig. 2.7 Impact of the
number of BS antennas N_t
on the performance of
Algorithm 2.1

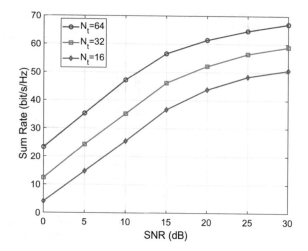

Next, we show the effect of the number of UEs N_m in a cluster on the sum
rate. As seen in Fig. 2.6, the sum rate decreases as the number of UEs in a cluster
increases. This is because there are more residual intra-cluster interference after
SIC if the number of UEs in a cluster is large. However, a small number of UEs in a
cluster means that the BS should designed more transmit beams, resulting in a higher
computational complexity. Moreover, if there are a small number of UEs in a cluster,
it is difficult to support massive access with a finite number of BS antennas. Thus, it
is desired to choose a proper number of UEs in a cluster.

In Fig. 2.7, we examine the impact of the number of BS antennas N_t on the
sum rate of Algorithm 2.1. It is seen that the sum rate improves as the number of
BS antennas increases. The reason is that increasing the number of BS antenna can
provide more array gains for enhancing the performance. Moreover, since the system
is interference limited at high SNR, the sum rate will be asymptotically saturated as
the SNR increases. Thus, increasing the number of BS antennas is a feasible method
for improving the performance at high SNR.

Figure 2.8 shows the total transmit power consumption at the BS versus the
required harvested energy threshold q_0 under the non-linear and linear EH models,
where $\gamma_0 = 0.1$ dB. It can be seen that the total transmit power consumption at the
BS for the both EH models increases with the increment of q_0. When $q_0 = 23.2$ mW,
there is a intersection point between the two curves associate with the non-linear and
linear EH models. When $q_0 < 23.2$ mW, the system designed with the non-linear
EH model always requires less transmit power than that designed with the linear EH
model. When $q_0 > 23.2$ mW, the optimal transmit beam and PS ratio designed under
the linear EH model cannot satisfy the information rate and energy harvested constraints, which means that in this case the linear EH model is infeasible for practical
system.

Figure 2.9 presents the total transmit power consumption at the BS versus the
required minimum SINR γ_0 under the non-linear and linear EH models, where

Fig. 2.8 Total transmit power versus q_0

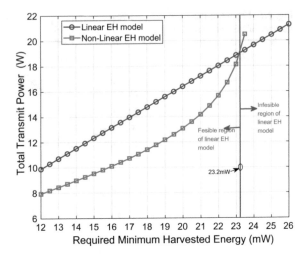

Fig. 2.9 Total transmit power versus γ_0

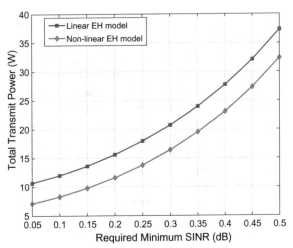

$q_0 = 20$ mW. It can be seen that the total transmit power improves as the required minimum SINR γ_0 increases, when the required minimum harvested energy q_0 remains unchanged. This is because more transmit power is required to meet the growing SINR threshold. Moreover, the wireless powered cellular IoT network designed with the non-linear EH model always consumes less transmit power than that designed with the linear EH model, which demonstrates that in practical system if the non-linear EH model is employed, a greater performance gain can be achieved compared with employing the linear one.

Finally, we check the impact of imperfect SIC on the power consumption of the proposed Algorithm 2.2 with $q_0 = 20$ mW in Fig. 2.10. It is intuitive that imperfect SIC will cause a high residual intra-cluster interference, resulting in a large transmit

Fig. 2.10 Effect of imperfect SIC on the TPC

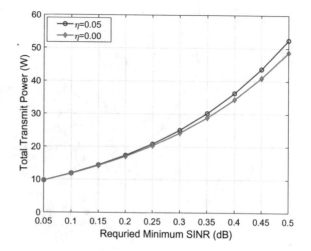

power consumption. However, it is found that there is a small gap of the TPC for the scenarios of $\eta = 0.05$ (imperfect SIC) and $\eta = 0.00$ (perfect SIC), especially when the required minimum SINR γ_0 is low. That is, the proposed Algorithm 2.2 has a strong capability for alleviating the impact of imperfect SIC.

2.3 Worst-Case Robust Design with Channel Quantization Error

As mentioned in Introduction, the BS may only obtain partial CSI through estimation or feedback, and thus there always exists channel uncertainty at the BS. In order to reduce feedback overhead, CSI needs to be quantized prior to feedback and thus leads to inevitable quantization errors, which are usually bounded in an ellipsoid. In this section, we provide worst-case robust design against channel uncertainty.

2.3.1 System Model

Let us consider a B5G cellular IoT network as shown in Fig. 2.11, where a BS equipped with N_t antennas simultaneously broadcasts information and energy to K IoT UEs equipped with single antenna over the same carrier. For the sake of increasing the efficiency of both information and power transfer, a multi-antenna BS adopting NOMA technique is considered where user clustering[2] and spatial beamforming are

---

[2]Carrying out user clustering in such a SWIPT system is beneficial to achieve a balance between the system performance and computational complexity. On the one hand, for the UEs in a cluster, the

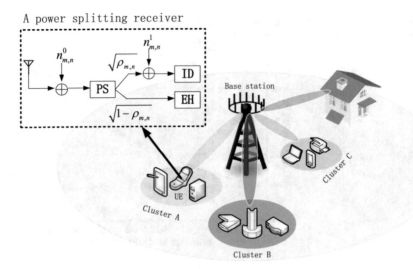

Fig. 2.11 A model of a cellular IoT network

applied. Specifically, the BS first obtains the spatial direction of each UE through some methods, like GPS or location tracking technique. Note that in this section, we consider a cellular massive IoT network where the UEs in general need to keep in touch with the multi-antenna BS in real time, and hence it is reasonably assumed that the BS can obtain the location of UEs. Then, the BS arranges UEs in the identical spatial direction but with distinct propagation distances into a cluster, and the UEs in the same cluster share a spatial beam. Without loss of generality, we assume that the K UEs are divided into M clusters, and the mth cluster, $\forall m \in \{1, \ldots, M\}$, contains N_m UEs. For simplicity of notation, we use $UE_{m,n}$ to represent the nth UE in the mth cluster. As shown in Fig. 2.11, a power splitting (PS) receiver is deployed at each UE, and thus the received RF signal at $UE_{m,n}$ is split into two components via PS with a $\rho_{m,n} : 1 - \rho_{m,n}$ proportion, where $0 \leq \rho_{m,n} \leq 1$ is the PS ratio. The $\rho_{m,n}$ part of the received RF signal is input to the ID receiver and the rest $1 - \rho_{m,n}$ part is sent to the EH receiver.

In the light of the available CSI, the BS implements superposition coding to realize massive spectrum sharing among the K IoT UEs. Generally speaking, superposition coding includes two steps for such a massive SWIPT system, i.e., power allocation and transmit beamforming. First, the BS constructs the transmit signal x_m for the mth cluster as below:

$$x_m = \sum_{n=1}^{N_m} \sqrt{\alpha_{m,n}} s_{m,n}, \tag{2.24}$$

---

identical spatial direction is conducive to boost the channel gain and to coordinate the inter-cluster interference. On the other hand, the distinctive propagation distances between the BS and UEs lessen the difficulty of the SIC at the IoT UEs.

where $s_{m,n}$ denotes the Gaussian distributed signal of unit norm for the $UE_{m,n}$, and $\alpha_{m,n}$ represents the intra-cluster power allocation factor for coordinating the intra-cluster interference. Then, the BS constructs the total transmit signal $\mathbf{x}$ as follows:

$$\mathbf{x} = \sum_{m=1}^{M} \mathbf{w}_m x_m, \tag{2.25}$$

where $\mathbf{w}_m$ is an N_t-dimensional transmit beam designed for the mth cluster based on available CSI for coordinating the inter-cluster interference. Finally, the BS broadcasts the signal $\mathbf{x}$ over the downlink channels. Therefore, the received signal at the $UE_{m,n}$ can be expressed as

$$y_{m,n} = \mathbf{h}_{m,n}^{H} \mathbf{x} + n_{m,n}^{0}, \tag{2.26}$$

where $\mathbf{h}_{m,n}$ denotes the N_t-dimensional channel vector from the BS to the $UE_{m,n}$. In fact, each UE knows its own effective channel gain by channel estimation for coherent detection. As a result, the BS can obtain the effective channel gains conveyed by UEs through the uplink channel and line them up in each cluster. Then, the order of effective channel gains is informed to the UEs via the downlink channel. Without loss of generality, we assume that the channel gains in the mth cluster are sorted in descending order as:

$$\left\| \mathbf{h}_{m,1} \right\|^2 \geq \left\| \mathbf{h}_{m,2} \right\|^2 \geq \cdots \geq \left\| \mathbf{h}_{m,N_m} \right\|^2. \tag{2.27}$$

Then, based on the order of channel gains in (2.27), the ith UE decodes and subtracts the interfering signals linked to the N_mth to $(i + 1)$th UEs at first. If SIC is carried out perfectly, the intra-cluster interference originated from the UEs with weaker channel gains can be removed entirely. However, as a matter of fact, the decoding error often occur at the weak interference signal because of the hardware restriction of the IoT UEs, the low signal quality and other factors. Thus, there exists residual interference from the weak UEs after SIC. As a rule, the residual interference resulting from imperfect SIC is an intractable function of multiple factors, e.g., coding/modulation related parameters, channel related issues (fading and shadowing), device/hardware/battery related restrictions, etc. Especially, owing to the traits of error propagation of imperfect SIC, it is hard to model the impact of imperfect SIC. Moreover, it is worth mentioning that for some certain simple IoT UEs with scanty computational capability, SIC is not a trivial task. That is to say, they do not have enough ability of carrying out SIC. In [17, 27], a linear model is given to capture the general impacts of SIC imperfection. Under such a model, via PS, the post-SIC signal at the $UE_{m,n}$ for ID can be expressed as

$$
y_{m,n}^{\mathrm{ID}} = \sqrt{\rho_{m,n}}(\mathbf{h}_{m,n}^{H}\mathbf{x} + n_{m,n}^{0}) + n_{m,n}^{1}
$$

$$
= \sqrt{\rho_{m,n}\alpha_{m,n}}\mathbf{h}_{m,n}^{H}\mathbf{w}_{m}s_{m,n} + \sqrt{\rho_{m,n}}\mathbf{h}_{m,n}^{H}\mathbf{w}_{m}\sum_{i=1}^{n-1}\sqrt{\alpha_{m,i}}s_{m,i} + \sqrt{\rho_{m,n}\eta_{m,n}}\mathbf{h}_{m,n}^{H}\mathbf{w}_{m}\sum_{i=n+1}^{N_{m}}\sqrt{\alpha_{m,i}}s_{m,i}
$$

$$
+ \sqrt{\rho_{m,n}}\mathbf{h}_{m,n}^{H}\sum_{j=1,j\neq m}^{M}\mathbf{w}_{j}\sum_{i=1}^{N_{j}}\sqrt{\alpha_{j,i}}s_{j,i} + \sqrt{\rho_{m,n}}n_{m,n}^{0} + n_{m,n}^{1}, \tag{2.28}
$$

where $n_{m,n}^{0}$ denotes AWGN at the $\mathrm{UE}_{m,n}$ with variance σ_{0}^{2}, $n_{m,n}^{1}$ is the baseband AWGN caused by the RF band to baseband signal conversion with variance σ_{1}^{2}, and $0 \leq \eta_{m,n} \leq 1$ represents the coefficient of imperfect SIC at the $\mathrm{UE}_{m,n}$, which can be acquired by long-term measurement.[3] Note that $\eta_{m,n} = 0$ means perfect SIC, $0 < \eta_{m,n} < 1$ represents imperfect SIC, and $\eta_{m,n} = 1$ denotes no SIC. As a consequence, the received SINR at the ID receiver of $\mathrm{UE}_{m,n}$ can be expressed as

$$
\Gamma_{m,n} = \frac{\alpha_{m,n}\left|\mathbf{h}_{m,n}^{H}\mathbf{w}_{m}\right|^{2}}{\left|\mathbf{h}_{m,n}^{H}\mathbf{w}_{m}\right|^{2}\sum_{i=1}^{n-1}\alpha_{m,i} + \eta_{m,n}\left|\mathbf{h}_{m,n}^{H}\mathbf{w}_{m}\right|^{2}\sum_{i=n+1}^{N_{m}}\alpha_{m,i} + \sum_{j=1,j\neq m}^{M}\left|\mathbf{h}_{m,n}^{H}\mathbf{w}_{j}\right|^{2} + \sigma_{0}^{2} + \frac{\sigma_{1}^{2}}{\rho_{m,n}}}. \tag{2.29}
$$

In addition, the received signal at the EH receiver of $\mathrm{UE}_{m,n}$ can be expressed as

$$
y_{m,n}^{\mathrm{EH}} = \sqrt{1 - \rho_{m,n}}(\mathbf{h}_{m,n}^{H}\mathbf{x} + n_{m,n}^{0}). \tag{2.30}
$$

According to (2.11), the input power at the EH receiver of $\mathrm{UE}_{m,n}$ is given by

$$
P_{m,n}^{\mathrm{in}} = (1 - \rho_{m,n})\left(\sum_{j=1}^{M}\left|\mathbf{h}_{m,n}^{H}\mathbf{w}_{j}\right|^{2} + \sigma_{0}^{2}\right). \tag{2.31}
$$

By adopting the non-linear EH model [30], the harvested energy at the $\mathrm{UE}_{m,n}$ is given by

$$
P_{m,n}^{(\mathrm{NL})} = \frac{\frac{M_{m,n}}{1+\exp(-a_{m,n}(P_{m,n}^{\mathrm{in}}-b_{m,n}))} - \frac{M_{m,n}}{1+\exp(a_{m,n}b_{m,n})}}{1 - \frac{1}{1+\exp(a_{m,n}b_{m,n})}}, \tag{2.32}
$$

where $M_{m,n}$, $a_{m,n}$ and $b_{m,n}$ are constants related to system parameters of the EH circuit. $M_{m,n}$ denotes the maximum harvested power at the $\mathrm{UE}_{m,n}$ with the saturated EH circuit, $a_{m,n}$ and $b_{m,n}$ are parameters in regard to the detailed circuit specification including the capacitance and resistance.

---

[3]Through off-line measuring a great number of samples for a long training time, based on the central limit theorem, the residual interference can be accurately approximated utilizing a Gaussian distribution, and the variance can be represented by the received power [43]. Then, the BS can obtain the imperfect SIC coefficient $\eta_{m,n}$ via comparing the powers of the residual interference and the received signal.

As seen in (2.29) and (2.32), the performance of both ID and EH is affected by the interference due to massive access. On the one hand, the interference decreases the quality of the received signal for ID. On the other hand, the interference increases the total amount of power of the received signal for EH. Thus, it is imperative to achieve a balance between ID and EH. Since the interference can be coordinated through transmit beamforming, optimizing the spatial beams serves as a key to improve the overall performance of the cellular IoT with SWIPT, especially under the adverse conditions of imperfect SIC and non-linear EH. Note that the optimization of spatial beams requires accurate CSI at the BS. However, in practical cellular IoT, the BS only has partial CSI. In this context, it is necessary to design robust beamforming algorithms for realizing sustainable communications of a massive number of IoT UEs.

2.3.2 Problem Formulation and Optimization Design

In this section, we design worst-case robust algorithms for the convergence of energy and communication in B5G cellular IoT with imperfect SIC and non-linear EH. Here, we adopt a deterministic CSI error model with the uncertainty assumed to lie in a given ellipsoid [19]. In particular, the relationship between the real channel vector $\mathbf{h}_{m,n}$ and the estimated CSI $\hat{\mathbf{h}}_{m,n}$ can be expressed as

$$\mathbf{h}_{m,n} \in \mathscr{H}_{m,n} = \{\hat{\mathbf{h}}_{m,n} + \mathbf{e}_{m,n} | \left\| \mathbf{e}_{m,n} \right\| \leq \varepsilon_{m,n}\}, \tag{2.33}$$

where $\mathbf{e}_{m,n}$ is the channel estimation error, the norm of which is bounded by $\varepsilon_{m,n}$. According to the characteristics of the cellular IoT, the WSR and the TPC are two mainly concerned performance metrics. Therefore, we design the robust algorithms from the perspectives of maximizing the WSR and minimizing the TPC, respectively.

2.3.2.1 The Weighted Sum Rate Maximization Design

With the above deterministic model of channel uncertainty in (2.33), we first design a robust algorithm for maximizing the WSR of the UEs while guaranteeing the EH requirements in the worst case. The robust design is formulated as the following optimization problem:

$$\max_{\mathbf{w}} \sum_{m=1}^{M} \sum_{n=1}^{N_m} \theta_{m,n} \min_{\mathbf{h}_{m,n} \in \mathscr{H}_{m,n}} R_{m,n}, \tag{2.34a}$$

$$\text{s.t.} \sum_{m=1}^{M} \left\| \mathbf{w}_m \right\|^2 \leq P_{\max}, \tag{2.34b}$$

$$\min_{\mathbf{h}_{m,n} \in \mathscr{H}_{m,n}} P_{m,n}^{(\text{NL})} \geq q_{m,n}, \tag{2.34c}$$

where $R_{m,n} = \log_2(1 + \Gamma_{m,n})$ is the achievable rate (in bit/s) of the $UE_{m,n}$ in the worst case, $\theta_{m,n} > 0$ denotes the priority of $UE_{m,n}$, $P_{\max} > 0$ is the maximum transmit power budget at the BS, and $q_{m,n} > 0$ is the required minimum amount of harvested energy, respectively. $\mathbf{w} = \{\mathbf{w}_1, \ldots, \mathbf{w}_M\}$ is the collection of spatial beams. Note that (2.34b) represents the maximum power allocation constraint, and constraint (2.34c) is the minimum requirement on the harvested energy of each UE. For convenience of expression, we define $Q_{m,n} \triangleq b_{m,n} - \frac{1}{a_{m,n}} \ln \frac{e^{a_{m,n}b_{m,n}}(M_{m,n}-q_{m,n})}{q_{m,n}e^{a_{m,n}b_{m,n}}+M_{m,n}}$, then constraint (2.34c) can be rewritten as

$$\min_{\mathbf{h}_{m,n} \in \mathcal{H}_{m,n}} \sum_{j=1}^{M} \left| \left(\hat{\mathbf{h}}_{m,n} + \mathbf{e}_{m,n} \right)^H \mathbf{w}_j \right|^2 + \sigma_0^2 \geq \frac{Q_{m,n}}{1 - \rho_{m,n}}. \tag{2.35}$$

It is clear that the problem (2.34) is not convex, thereby it is difficult to obtain the optimal solution directly. Especially, the BS only has partial CSI, which further leads to a complicated objective function. In general, a brute force approach is required to obtain the globally solution of the problem at hand. However, it will incur a prohibitively large complicated complexity. To deal with this issue, we resort to optimizing a lower bound on the objective function. To this end, we first introduce the following lemma.

Lemma 2.2 *For two optimization problems as follow:*

$$f_1(\mathbf{x}) = \max_{\|\mathbf{x}\| \leq \varepsilon} \text{Re}\{\mathbf{x}^H \mathbf{y}\}, \tag{2.36a}$$

and

$$f_2(\mathbf{x}) = \min_{\|\mathbf{x}\| \leq \varepsilon} \text{Re}\{\mathbf{x}^H \mathbf{y}\}, \tag{2.36b}$$

where variable $\mathbf{x}$ is norm-bounded with ε and $\mathbf{y}$ is a given parameter. Their solutions are given by

$$f_1\left(\frac{\varepsilon}{\|\mathbf{y}\|}\mathbf{y}\right) = \varepsilon \|\mathbf{y}\| \tag{2.37a}$$

and

$$f_2\left(-\frac{\varepsilon}{\|\mathbf{y}\|}\mathbf{y}\right) = -\varepsilon \|\mathbf{y}\|, \tag{2.37b}$$

respectively.

Proof It is known that $-\left|\mathbf{x}^H\mathbf{y}\right| \leq \text{Re}\{\mathbf{x}^H\mathbf{y}\} \leq \left|\mathbf{x}^H\mathbf{y}\right|$. Then, according to the Cauchy-Schwarz inequality [37], the equation $|\langle \mathbf{x}, \mathbf{y}\rangle|^2 \leq \langle \mathbf{x}, \mathbf{x}\rangle \cdot \langle \mathbf{y}, \mathbf{y}\rangle$ holds true if and only if vector $\mathbf{x}$ and vector $\mathbf{y}$ are aligned, where $\langle \cdot, \cdot\rangle$ denotes the vector inner product. Thus, we have

$$\left|\mathbf{x}^H\mathbf{y}\right| \leq \|\mathbf{x}\| \|\mathbf{y}\| \leq \varepsilon \|\mathbf{y}\|, \tag{2.38}$$

and

$$-\varepsilon \|\mathbf{y}\| \leq \mathrm{Re}\{\mathbf{x}^H \mathbf{y}\} \leq \varepsilon \|\mathbf{y}\|. \tag{2.39}$$

In particular, the upper bound of $\mathrm{Re}\{\mathbf{x}^H \mathbf{y}\}$ is $\varepsilon \|\mathbf{y}\|$ at the point $\mathbf{x} = \frac{\varepsilon}{\|\mathbf{y}\|}\mathbf{y}$, and the lower bound of $\mathrm{Re}\{\mathbf{x}^H \mathbf{y}\}$ is $-\varepsilon \|\mathbf{y}\|$ at the point $\mathbf{x} = -\frac{\varepsilon}{\|\mathbf{y}\|}\mathbf{y}$. $\qquad \square$

Then, we deal with the term $\left|\mathbf{h}_{m,n}^H \mathbf{w}_m\right|^2$ in the objective function (2.34a) by utilizing Lemma 2.2. Specifically, we have

$$
\begin{aligned}
\left|\mathbf{h}_{m,n}^H \mathbf{w}_m\right|^2 &= (\hat{\mathbf{h}}_{m,n} + \mathbf{e}_{m,n})^H \mathbf{w}_m \mathbf{w}_m^H (\hat{\mathbf{h}}_{m,n} + \mathbf{e}_{m,n}) \\
&= \hat{\mathbf{h}}_{m,n}^H \mathbf{w}_m \mathbf{w}_m^H \hat{\mathbf{h}}_{m,n} + 2\mathrm{Re}\{\mathbf{e}_{m,n}^H \mathbf{w}_m \mathbf{w}_m^H \hat{\mathbf{h}}_{m,n}\} + \mathbf{e}_{m,n}^H \mathbf{w}_m \mathbf{w}_m^H \mathbf{e}_{m,n} \\
&\geq \hat{\mathbf{h}}_{m,n}^H \mathbf{w}_m \mathbf{w}_m^H \hat{\mathbf{h}}_{m,n} + 2\mathrm{Re}\{\mathbf{e}_{m,n}^H \mathbf{w}_m \mathbf{w}_m^H \hat{\mathbf{h}}_{m,n}\},
\end{aligned} \tag{2.40}
$$

where in (2.40) the second-order error term $\mathbf{e}_{m,n}^H \mathbf{w}_m \mathbf{w}_m^H \mathbf{e}_{m,n}$ is ignored due to its tiny value compared to other terms. Based on Lemma 2.2, we can obtain the upper bound and the lower bound of $\left|\mathbf{h}_{m,n}^H \mathbf{w}_m\right|^2$ as

$$\left|\mathbf{h}_{m,n}^H \mathbf{w}_m\right|^2 \leq \hat{\mathbf{h}}_{m,n}^H \mathbf{w}_m \mathbf{w}_m^H \hat{\mathbf{h}}_{m,n} + 2\varepsilon_{m,n} \left\|\mathbf{w}_m \mathbf{w}_m^H \hat{\mathbf{h}}_{m,n}\right\| \tag{2.41}$$

and

$$\left|\mathbf{h}_{m,n}^H \mathbf{w}_m\right|^2 \geq \hat{\mathbf{h}}_{m,n}^H \mathbf{w}_m \mathbf{w}_m^H \hat{\mathbf{h}}_{m,n} - 2\varepsilon_{m,n} \left\|\mathbf{w}_m \mathbf{w}_m^H \hat{\mathbf{h}}_{m,n}\right\|, \tag{2.42}$$

respectively. Thus, the lower bound of the objective function in (2.34) can be written as

$$\sum_{m=1}^{M}\sum_{n=1}^{N_m} \log_2\left(\frac{\sum_{j=1}^{M}\sum_{i=1}^{N_j} \phi_{j,i}^{m,n}\alpha_{j,i}\left(\hat{\mathbf{h}}_{m,n}^H \mathbf{w}_j \mathbf{w}_j^H \hat{\mathbf{h}}_{m,n} - 2\varepsilon_{m,n}\left\|\mathbf{w}_j \mathbf{w}_j^H \hat{\mathbf{h}}_{m,n}\right\|\right) + \sigma_0^2 + \frac{\sigma_1^2}{\rho_{m,n}}}{\sum_{j=1}^{M}\sum_{i=1}^{N_j} \varphi_{j,i}^{m,n}\alpha_{j,i}\left(\hat{\mathbf{h}}_{m,n}^H \mathbf{w}_j \mathbf{w}_j^H \hat{\mathbf{h}}_{m,n} + 2\varepsilon_{m,n}\left\|\mathbf{w}_j \mathbf{w}_j^H \hat{\mathbf{h}}_{m,n}\right\|\right) + \sigma_0^2 + \frac{\sigma_1^2}{\rho_{m,n}}} \right), \tag{2.43}$$

where $\phi_{j,i}^{m,n}$ and $\varphi_{j,i}^{m,n}$ are auxiliary variables for convenience of expression, which are defined as

$$
\phi_{j,i}^{m,n} = \begin{cases} \eta_{m,n}, & \text{if } j = m \text{ and } i > n \\ 0, & \text{if } j = m \text{ and } i = n \\ 1, & \text{otherwise} \end{cases} \quad \text{and} \quad \varphi_{j,i}^{m,n} = \begin{cases} \eta_{m,n}, & \text{if } j = m \text{ and } i > n \\ 1, & \text{otherwise} \end{cases}. \tag{2.44}
$$

To make the problem more tractable, we define $\mathbf{W}_m = \mathbf{w}_m \mathbf{w}_m^H$ and let

$$\zeta_{m,n} = \sum_{j=1}^{M}\sum_{i=1}^{N_j} \phi_{j,i}^{m,n}\alpha_{j,i}\left(\hat{\mathbf{h}}_{m,n}^H \mathbf{W}_j \hat{\mathbf{h}}_{m,n} - 2\varepsilon_{m,n}\left\|\mathbf{W}_j \hat{\mathbf{h}}_{m,n}\right\|\right) + \sigma_0^2 + \frac{\sigma_1^2}{\rho_{m,n}} \tag{2.45}$$

and

$$\varpi_{m,n} = \sum_{j=1}^{M} \sum_{i=1}^{N_j} \varphi_{j,i}^{m,n} \alpha_{j,i} \left(\hat{\mathbf{h}}_{m,n}^{H} \mathbf{W}_j \hat{\mathbf{h}}_{m,n} + 2\varepsilon_{m,n} \left\| \mathbf{W}_j \hat{\mathbf{h}}_{m,n} \right\| \right) + \sigma_0^2 + \frac{\sigma_1^2}{\rho_{m,n}}. \quad (2.46)$$

Thus, the objective function can be rewritten as

$$\max_{\mathbf{W}} \log_2 \prod_{m=1}^{M} \prod_{n=1}^{N_m} \frac{\zeta_{m,n}}{\varpi_{m,n}}, \quad (2.47)$$

where $\mathbf{W} = \{\mathbf{W}_1, \mathbf{W}_2, \cdots, \mathbf{W}_M\}$ is the collection of new introduced matrices. In order to further deal with the non-convex objective function (2.47), we define

$$e^{x_{m,n}} \triangleq \zeta_{m,n} \text{ and } e^{y_{m,n}} \triangleq \varpi_{m,n}, \quad (2.48)$$

where $x_{m,n}$ and $y_{m,n}$ are introduced slack variables. Since the upper bound and the lower bound of $\left| \mathbf{h}_{m,n}^H \mathbf{x} \right|^2$ are both nonnegative, we have

$$e^{x_{m,n}} \geq \sigma_0^2 + \frac{\sigma_1^2}{\rho_{m,n}} \text{ and } e^{y_{m,n}} \geq \sigma_0^2 + \frac{\sigma_1^2}{\rho_{m,n}}. \quad (2.49)$$

Moreover, due to the constraint of transmit power at the BS, the slack variables $x_{m,n}$ and $y_{m,n}$ are finite. Substituting (2.48) into the objective function (2.47), we obtain

$$\max_{\mathbf{W}} \log_2 \prod_{m=1}^{M} \prod_{n=1}^{N_m} \frac{\zeta_{m,n}}{\varpi_{m,n}} = \max_{\mathbf{x},\mathbf{y},\mathbf{W}} \log_2 \prod_{m=1}^{M} \prod_{n=1}^{N_m} (e^{x_{m,n} - y_{m,n}})$$

$$= \max_{\mathbf{x},\mathbf{y},\mathbf{W}} \sum_{m=1}^{M} \sum_{n=1}^{N_m} (x_{m,n} - y_{m,n})\log_2 e, \quad (2.50)$$

where $\mathbf{x} = \{x_{1,1}, \ldots, x_{M,N_m}\}$ and $\mathbf{y} = \{y_{1,1}, \ldots, y_{M,N_m}\}$ are the collections of slack variables. Meanwhile, dropping the rank-one constraint, i.e., $\text{Rank}(\mathbf{W}_m) = 1$, problem (2.34) can be reformulated as the following SDP problem:

$$\max_{\mathbf{x},\mathbf{y},\mathbf{W}} \sum_{m=1}^{M} \sum_{n=1}^{N_m} (x_{m,n} - y_{m,n})\log_2 e \quad (2.51a)$$

$$\text{s.t. } (2.49),$$

$$\sum_{m=1}^{M} \text{tr}(\mathbf{W}_m) \leq P_{\max}, \quad (2.51b)$$

$$\sum_{j=1}^{M} \left(\hat{\mathbf{h}}_{m,n}^{H} \mathbf{W}_j \hat{\mathbf{h}}_{m,n} - 2\varepsilon_{m,n} \left\| \mathbf{W}_j \hat{\mathbf{h}}_{m,n} \right\| \right) + \sigma_0^2 \geq \frac{Q_{m,n}}{1 - \rho_{m,n}}, \quad (2.51c)$$

$$\mathbf{W}_m \succeq \mathbf{0}, \forall m, \quad (2.51d)$$

$$x_{m,n} \geq y_{m,n}, \quad (2.51e)$$

$$\zeta_{m,n} \geq e^{x_{m,n}}, \quad (2.51f)$$

$$\varpi_{m,n} \leq e^{y_{m,n}}. \quad (2.51g)$$

Although the objective function (2.51a) and constraint (2.51f) have been transformed to convex ones, the constraint (2.51g) is still non-convex. In order to solve this issue, a Taylor series expansion $f(x) = \sum_{n=0}^{\infty} \frac{f^{(n)}}{n!}(x - x_0)^n + R_n(x)$ is applied. Let us define

$$\tilde{y}_{m,n} \triangleq \ln \left(\sum_{j=1}^{M} \sum_{i=1}^{N_j} \varphi_{j,i}^{m,n} \alpha_{j,i} \left(\hat{\mathbf{h}}_{m,n}^{H} \tilde{\mathbf{W}}_j \hat{\mathbf{h}}_{m,n} + 2\varepsilon_{m,n} \left\| \tilde{\mathbf{W}}_j \hat{\mathbf{h}}_{m,n} \right\| \right) + \sigma_0^2 + \frac{\sigma_1^2}{\rho_{m,n}} \right),$$
$$(2.52)$$

where $\tilde{\mathbf{W}}_m$ is a feasible point of problem (2.51) at the last iteration. Since $e^{y_{m,n}}$ is convex, its first-order Taylor series expansion at $\tilde{y}_{m,n}$ is $e^{\tilde{y}_{m,n}} \left(y_{m,n} - \tilde{y}_{m,n} + 1 \right)$. Thus, the constraint (2.51g) can be transformed as

$$\varpi_{m,n} \leq e^{\tilde{y}_{m,n}} \left(y_{m,n} - \tilde{y}_{m,n} + 1 \right). \quad (2.53)$$

Then, for a given $\tilde{\mathbf{W}}_m$, problem (2.51) can be rewritten as

$$\max_{\mathbf{x},\mathbf{y},\mathbf{W}} \quad \log_2 e \sum_{m=1}^{M} \sum_{n=1}^{N_m} (x_{m,n} - y_{m,n}) \quad (2.54a)$$

$$\text{s.t.} \quad (2.49), (2.51b)\text{-}(2.51f), (2.53).$$

Hence, problem (2.54) is a convex optimization problem which can be effectively solved by some optimization software, i.e., CVX [36]. Furthermore, by iteratively updating the fixed point $\tilde{\mathbf{W}}_m$ according to (2.52) and (2.53), we can obtain the final solutions $\mathbf{W}_m$, $\forall m$, once they converge. For this proposed iteration algorithm, we have the following theorem.

Theorem 2.1 *The solutions $\{x_{m,n}, y_{m,n}, \mathbf{W}_m\}$ to the problem (2.54) can be converged in the iterations.*

Proof Please refer to Appendix C. □

Note that we drop the rank-one constraint in problem (2.54). In order to recover the concerned spatial beam $\mathbf{w}_m$, the $\mathbf{W}_m$ should be rank-one. According to Pataki's result

([38], Theorem 2.2), an optimal solution $\mathbf{W}_m^*$ for the downlink transmit beamforming system with multiple unicast should satisfy

$$\sum_{m=1}^{M} \frac{\text{Rank}(\mathbf{W}_m^*)\left(\text{Rank}(\mathbf{W}_m^*) + 1\right)}{2} \leq M. \tag{2.55}$$

Since all the optimal solution $\mathbf{W}_m^* \neq \mathbf{0}$, we have $\text{Rank}(\mathbf{W}_m^*) \geq 1$. Combine with (2.55), it is easy to obtain $\text{Rank}(\mathbf{W}_m^*) = 1$, which means that the semi-definite relaxation (SDR) technique, i.e., dropping the rank-one constraint employed in problem (2.54), is tight. Therefore, we can obtain the unique solution $\mathbf{w}_m^*$ to problem (2.54) by eigenvalue decomposition (EVD) on $\mathbf{W}_m^*$, namely

$$\mathbf{w}_m^* = \sqrt{\lambda_{\max}\left(\mathbf{W}_m^*\right)}\mathbf{v}_{m,\max}^*, \tag{2.56}$$

where $\lambda_{\max}\left(\mathbf{W}_m^*\right)$ is the maximum eigenvalue of $\mathbf{W}_m^*$, and $\mathbf{v}_{m,\max}^*$ is the unit eigenvector with respect to $\lambda_{\max}\left(\mathbf{W}_m^*\right)$. In addition, for the first-order Taylor series expansion of $e^{y_{m,n}}$ at point $\tilde{y}_{m,n}$ in constraint (2.51g), the closer the expansion point $\tilde{y}_{m,n}$ is to the variable $y_{m,n}$, the smaller the approximation error is. Hence, in order to set feasible initial values $\tilde{y}_{m,n}^{(0)}$ to accelerate the convergence of algorithm, we make $\tilde{\mathbf{w}}_m^{(0)} = \sqrt{\frac{P_{\max}}{M}}[1, 0, \cdots, 0]^T$ to satisfy the transmission power constraint. Accordingly, the initial value of $\tilde{y}_{m,n}^{(0)}$ is computed by (2.52). In summary, a robust beamforming design with massive SWIPT for maximizing the WSR can be described as Algorithm 2.3.

Algorithm 2.3 Worst-Case Robust Design for Maximizing the WSR.

Input: $\theta_{m,n}, \eta_{m,n}, \alpha_{m,n}, \rho_{m,n}, q_{m,n}, \sigma_0^2, \sigma_1^2, \varepsilon_{m,n}, P_{\max}$.
Output: $\mathbf{w}_m$.

1: **Initialize** $\mathbf{w}_m^{(0)} = \sqrt{\frac{P_{\max}}{M}}[1, 0, \cdots, 0]^T, \forall m$, the convergence accuracy Δ, the iteration index

   $t = 1$, the WSR $R^{(0)} = 0$, and the feasible point $\tilde{\mathbf{W}}_m^{(0)} = \mathbf{w}_m^{(0)}(\mathbf{w}_m^{(0)})^H$;
2: **while** $\Delta > 0.01$ **do**
3:    Solve the convex problem (2.54) by CVX and get the optimal solution $\mathbf{W}_m^*$;
4:    Update $\tilde{\mathbf{W}}_m^{(t)} = \mathbf{W}_m^*$, $\Delta = R^{(t)} - R^{(t-1)}$;
5:    Update $t = t + 1$;
6: **end while**
7: Obtain $\mathbf{w}_m^*$ by EVD on $\mathbf{W}_m^*$ according to (2.56);

Remark 2.2 To explicitly show the impact of the capacity of IoT UEs on the performance of the proposed algorithm, we use the parameters $\phi_{j,i}^{m,n}$ and $\varphi_{j,i}^{m,n}$ to show the impact of imperfect SIC, i.e., $\phi_{j,i}^{m,n} = \psi_{j,i}^{m,n} = \eta_{m,n}, \forall j = m$ and $i > n$. Specifically, $\eta_{m,n} = 1$ represents that the IoT UE has no capability to perform SIC, $\eta_{m,n} = 0$ represents the ideal situation of perfect SIC, and if the IoT UE performs SIC but SIC is imperfect, $\eta_{m,n}$ is within the range (0, 1). In addition, if the channel estimation error bound $\varepsilon_{m,n} = 0$, it becomes the case of full CSI. Therefore, Algorithm 2.3 provides a general design method for massive SWIPT in cellular IoT.

2.3.2.2 The Total Power Consumption Minimization Design

In this section, we design a robust massive SWIPT algorithm for the cellular IoT network with the goal of minimizing the TPC at the BS, while guaranteeing QoS requirements. It is worth pointing out that although the IoT applications may have different QoS requirements, e.g., the delay, throughput and reliability, they all can be represented as a function of SINR. Thereby, the optimization problem related to the minimization of the TPC can be mathematically expressed as

$$\min_{\mathbf{w}} \sum_{m=1}^{M} \|\mathbf{w}_m\|^2 \tag{2.57a}$$

$$\text{s.t.} \quad \min_{\mathbf{h}_{m,n} \in \mathscr{H}_{m,n}} \frac{\alpha_{m,n} |\mathbf{h}_{m,n}^H \mathbf{w}_m|^2}{\sum_{j=1}^{M} \sum_{i=1}^{N_j} \phi_{j,i}^{m,n} \alpha_{j,i} |\mathbf{h}_{m,n}^H \mathbf{w}_j|^2 + \sigma_0^2 + \frac{\sigma_1^2}{\rho_{m,n}}} \geq \gamma_{m,n}, \tag{2.57b}$$

$$\min_{\mathbf{h}_{m,n} \in \mathscr{H}_{m,n}} \sum_{j=1}^{M} |\mathbf{h}_{m,n}^H \mathbf{w}_j|^2 + \sigma_0^2 \geq \frac{Q_{m,n}}{1 - \rho_{m,n}}, \tag{2.57c}$$

where $\gamma_{m,n} > 0$ is the prescribed SINR target for $UE_{m,n}$, $\mathbf{w} = \{\mathbf{w}_1, \ldots, \mathbf{w}_M\}$ is the collection of spatial beams. It is seen that problem (2.57) is not convex due to infinitely many quadratic inequalities for SINR and EH constraints. To make problem (2.57) computationally tractable, we first utilize new matrix variables $\mathbf{W}_m = \mathbf{w}_m \mathbf{w}_m^H$ and introduce two useful lemmas as follows.

Lemma 2.3 *(S-procedure, [35]) Let us consider a function* $\mathbf{f}_m(\mathbf{x})$ *as*

$$\mathbf{f}_m(\mathbf{x}) = \mathbf{x}^H \mathbf{A}_m \mathbf{x} + 2\operatorname{Re}\{\mathbf{b}_m^H \mathbf{x}\} + c_m, \ m \in \{1, 2\}, \mathbf{x} \in \mathbb{C}^{N \times 1}, \tag{2.58}$$

where $\mathbf{A}_m \in \mathbb{C}^{N \times N}, \mathbf{b}_m \in \mathbb{C}^{N \times 1}$ *and* $c_m \in \mathbb{C}^{N \times 1}$. *The derivation* $\mathbf{f}_1(\mathbf{x}) \leq 0 \Rightarrow \mathbf{f}_2(\mathbf{x}) \leq 0$ *holds true if and only if there exists* $\tau \geq 0$, *such that*

$$\tau \begin{bmatrix} \mathbf{A}_1 & \mathbf{b}_1 \\ \mathbf{b}_1^H & c_1 \end{bmatrix} - \begin{bmatrix} \mathbf{A}_2 & \mathbf{b}_2 \\ \mathbf{b}_2^H & c_2 \end{bmatrix} \succeq \mathbf{0}. \tag{2.59}$$

Lemma 2.4 *(Schur's complement, [35]) Let* $\mathbf{Y} = \begin{bmatrix} \mathbf{A} & \mathbf{B}^H \\ \mathbf{B} & \mathbf{C} \end{bmatrix}$ *be a Hermitian matrix. Then* $\mathbf{Y} \succeq \mathbf{0}$ *holds true if and only if* $\mathbf{A} - \mathbf{B}^H \mathbf{C}^{-1} \mathbf{B} \succeq \mathbf{0}$ *with assuming* $\mathbf{C}$ *is invertible, or* $\mathbf{C} - \mathbf{B}^H \mathbf{A}^{-1} \mathbf{B} \succeq \mathbf{0}$ *with assuming* $\mathbf{A}$ *is invertible.*

Now, we deal with the SINR and EH constraints based on above two lemmas. For the SINR constraint of the $UE_{m,n}$, it can be expressed as

$$\rho_{m,n}\left[\left(\hat{\mathbf{h}}_{m,n}+\mathbf{e}_{m,n}\right)^{H}\left(\frac{1}{\gamma_{m,n}}\mathbf{W}_m-\sum_{j=1}^{M}\sum_{i=1}^{N_j}\phi_{j,i}^{m,n}\alpha_{j,i}\mathbf{W}_j\right)\left(\hat{\mathbf{h}}_{m,n}+\mathbf{e}_{m,n}\right)-\sigma_0^2\right]\geq\sigma_1^2.$$

(2.60)

Let $\mathbf{Y}_{m,n}=\frac{1}{\gamma_{m,n}}\mathbf{W}_m-\sum_{j=1}^{M}\sum_{i=1}^{N_j}\phi_{j,i}^{m,n}\alpha_{j,i}\mathbf{W}_j$, the inequality (2.60) can be rewritten as

$$\left(\hat{\mathbf{h}}_{m,n}+\mathbf{e}_{m,n}\right)^{H}\mathbf{Y}_{m,n}\left(\hat{\mathbf{h}}_{m,n}+\mathbf{e}_{m,n}\right)-\sigma_0^2-\frac{\sigma_1^2}{\rho_{m,n}}\geq 0,$$

(2.61)

which is equivalent to the following inequality

$$\mathbf{e}_{m,n}^{H}\mathbf{Y}_{m,n}\mathbf{e}_{m,n}+2\operatorname{Re}\left\{\hat{\mathbf{h}}_{m,n}^{H}\mathbf{Y}_{m,n}\mathbf{e}_{m,n}\right\}+\hat{\mathbf{h}}_{m,n}^{H}\mathbf{Y}_{m,n}\hat{\mathbf{h}}_{m,n}-\left(\sigma_0^2+\frac{\sigma_1^2}{\rho_{m,n}}\right)\geq 0.$$

(2.62)

Applying Lemmas 2.3 and 2.4, and setting $\mathbf{x}=\mathbf{e}_{m,n}$ and $\tau=\beta_{m,n}$, we have

$$\begin{bmatrix}-\mathbf{Y}_{m,n} & -\mathbf{Y}_{m,n}\hat{\mathbf{h}}_{m,n} \\ -\hat{\mathbf{h}}_{m,n}^{H}\mathbf{Y}_{m,n}^{H} & \left(\sigma_0^2+\frac{\sigma_1^2}{\rho_{m,n}}\right)-\hat{\mathbf{h}}_{m,n}^{H}\mathbf{Y}_{m,n}\hat{\mathbf{h}}_{m,n}\end{bmatrix}\preceq\beta_{m,n}\begin{bmatrix}\mathbf{I} & \mathbf{0} \\ \mathbf{0} & -\varepsilon_{m,n}^2\end{bmatrix}.$$

(2.63)

Thus, the $UE_{m,n}$'s SINR constraint (2.57b) can be transformed as a linear matrix inequality (LMI) constraint, i.e.,

$$\mathbf{A}_{m,n}=\begin{bmatrix}\beta_{m,n}\mathbf{I}+\mathbf{Y}_{m,n} & \mathbf{Y}_{m,n}\hat{\mathbf{h}}_{m,n} \\ \hat{\mathbf{h}}_{m,n}^{H}\mathbf{Y}_{m,n}^{H} & \hat{\mathbf{h}}_{m,n}^{H}\mathbf{Y}_{m,n}\hat{\mathbf{h}}_{m,n}+\upsilon_{m,n}\end{bmatrix}\succeq\mathbf{0},$$

(2.64)

where $\upsilon_{m,n}=-\beta_{m,n}\varepsilon_{m,n}^2-\sigma_0^2-\frac{\sigma_1^2}{\rho_{m,n}}$. Similarly, we can obtain an equivalent form of the EH constraint (2.57c) as

$$\mathbf{B}_{m,n}=\begin{bmatrix}\mu_{m,n}\mathbf{I}+\mathbf{T} & \mathbf{T}\hat{\mathbf{h}}_{m,n} \\ \hat{\mathbf{h}}_{m,n}^{H}\mathbf{T}^{H} & \hat{\mathbf{h}}_{m,n}^{H}\mathbf{T}\hat{\mathbf{h}}_{m,n}+\varsigma_{m,n}\end{bmatrix}\succeq\mathbf{0},$$

(2.65)

where $\mathbf{T}=\sum_{j=1}^{M}\mathbf{W}_j$ and $\varsigma_{m,n}=-\mu_{m,n}\varepsilon_{m,n}^2+\sigma_0^2-\frac{Q_{m,n}}{1-\rho_{m,n}}$. Hence, we can reformulate problem (2.57) with transformed SINR constraint (2.64) and EH constraint (2.65) as

$$\min_{\mathbf{W},\boldsymbol{\beta},\boldsymbol{\mu}} \sum_{m=1}^{M} \text{tr}\,(\mathbf{W}_m) \tag{2.66a}$$

$$\text{s.t.}\quad (2.64),\ (2.65),$$

$$\beta_{m,n} \geq 0,\ \mu_{m,n} \geq 0, \tag{2.66b}$$

$$\mathbf{W}_m \succeq \mathbf{0}, \tag{2.66c}$$

$$\text{Rank}(\mathbf{W}_m) = 1, \tag{2.66d}$$

where $\boldsymbol{\beta} = \{\beta_{1,1}, \ldots, \beta_{M,N}\}$ and $\boldsymbol{\mu} = \{\mu_{1,1}, \ldots, \mu_{M,N}\}$ are the collections of above LMIs' auxiliary coefficients, respectively. It is obvious that if we adopt a general SDR technique, i.e., dropping the rank-one constraint (2.66d), the problem (2.66) is convex and can be efficiently solved via an off-the-shelf optimization tool, e.g., CVX. However, the solution $\mathbf{W}_m$ obtained by the SDR technique is not always rank-one. In this case, we utilize an alternative method based on a penalty function to guarantee $\mathbf{W}_m$ is rank-one [35]. Since $\mathbf{W}_m$ is positive semi-definite according to (2.66c), all the eigenvalues of $\mathbf{W}_m$ are non-negative, i.e., $\lambda_i(\mathbf{W}_m) \geq 0, \forall m, i = 1, 2, \ldots, N_t$. Due to the fact that $\text{tr}\,(\mathbf{W}_m) = \sum_{i=1}^{N_t} \lambda_i(\mathbf{W}_m)$, we have $\text{tr}(\mathbf{W}_m) \geq \lambda_{\max}(\mathbf{W}_m)$, where $\lambda_{\max}(\cdot)$ denotes the maximum eigenvalue of a matrix. Thus, for a positive semi-definite matrix $\mathbf{W}_m$ with the constraint of $\text{tr}(\mathbf{W}_m) = \lambda_{\max}(\mathbf{W}_m)$, it is easy to derive $\text{Rank}(\mathbf{W}_m) = 1$. In other words, the rank-one constraint (2.66d) can be rewritten as

$$\text{tr}(\mathbf{W}_m) - \lambda_{\max}(\mathbf{W}_m) = 0, \forall m. \tag{2.67}$$

Inspired by (2.67), we argument the difference of $\text{tr}(\mathbf{W}_m)$ and $\lambda_{\max}(\mathbf{W}_m)$ to the objective function as the penalty function. Through minimizing the penalty function, it is likely to fulfill the rank-one condition in (2.67). Hence, the new objective function of problem (2.66) can be expressed as

$$\min_{\mathbf{W},\boldsymbol{\beta},\boldsymbol{\mu}} \sum_{m=1}^{M} \text{tr}(\mathbf{W}_m) + \kappa \sum_{m=1}^{M} (\text{tr}(\mathbf{W}_m) - \lambda_{\max}(\mathbf{W}_m)), \tag{2.68}$$

where $\kappa > 0$ is the penalty factor. However, the introduction of the penalty function leads to a non-convex objective function. To obtain a tractable solution, we adopt an iterative method to transform it into a convex one. Specifically, for the feasible point $\mathbf{W}_m^{(t)}$ at the tth iteration, we have

$$\text{tr}\left(\mathbf{W}_m^{(t+1)}\right) - \left(\mathbf{v}_{m,\max}^{(t)}\right)^H \mathbf{W}_m^{(t+1)} \mathbf{v}_{m,\max}^{(t)} \geq \text{tr}\left(\mathbf{W}_m^{(t+1)}\right) - \lambda_{\max}\left(\mathbf{W}_m^{(t+1)}\right) \geq 0, \tag{2.69}$$

where $\mathbf{v}_{m,\max}$ is the unit eigenvector with respect to the maximum eigenvalue $\lambda_{\max}(\mathbf{W}_m)$. Thus, the problem (2.66) at $(t+1)$th iteration is given by

$$\min_{\mathbf{W},\mu,\beta} \sum_{m=1}^{M} \text{tr}(\mathbf{W}_m^{(t+1)}) + \kappa^{(t+1)} \sum_{m=1}^{M} \left(\text{tr}(\mathbf{W}_m^{(t+1)}) - \left(\mathbf{v}_{m,\max}^{(t)}\right)^H \mathbf{W}_m^{(t+1)} \mathbf{v}_{m,\max}^{(t)} \right) \quad (2.70)$$

s.t. (2.64), (2.65), (2.66b), and (2.66c).

Note that problem (2.70) is convex, and thus can be solved directly. Since the penalty factor $\kappa^{(t+1)} = c\kappa^{(t)} (c > 1)$ increases during the iterations, the value of the penalty function is correspondingly decreasing. Moreover, the penalty function has a lower bound according to (2.69). Therefore, through iteratively updating the penalty factor κ, the objective function can be converged. Once the iteration converges, namely when $\text{tr}(\mathbf{W}_m)$ is approximately equal to $\lambda_{\max}(\mathbf{W}_m)$, we have

$$\mathbf{W}_m \approx \lambda_{\max}(\mathbf{W}_m)\mathbf{v}_{m,\max}\mathbf{v}_{m,\max}^H, \quad (2.71)$$

Accordingly, the approximate solution to the original problem (2.57) can be obtained as

$$\mathbf{w}_m^* = \sqrt{\lambda_{\max}(\mathbf{W}_m)}\mathbf{v}_{m,\max}. \quad (2.72)$$

In summary, the proposed penalty function-based iterative method can be described as Algorithm 2.4.

Algorithm 2.4 Worst-Case Robust Design for Minimizing the TPC

Input: $a_{m,n}, b_{m,n}, M_{m,n}, \eta_{m,n}, \alpha_{m,n}, \rho_{m,n}, \sigma_0^2, \sigma_1^2, q_{m,n}, \gamma_{m,n}$
Output: $\mathbf{w}_m$
1: **Initialize** $\mathbf{w}_m^{(0)} = \sqrt{\frac{P_{\max}}{M}}[1, 0, \cdots, 0]^T, \mathbf{W}_m^{(0)} = \mathbf{w}_m^{(0)}(\mathbf{w}_m^{(0)})^H$, a proper penalty factor κ, a proper coefficient $c > 1$, the convergence accuracy $\Delta = 10^{-12}$, the maximum number of iteration $T_{\max} = 30$ and the iteration number $t = 1$.
2: **repeat**
3:    Solve problem (2.66) with CVX, then obtain $\mathbf{W}_m^{(t)}$.
4:    **if** $\mathbf{W}_m^{(t)}$ converges **then**
5:       **if** $\sum_{m=1}^{M} | \text{tr}(\mathbf{W}_m^{(t)}) - \lambda_{\max}(\mathbf{W}_m^{(t)}) | > \Delta$ **then**
6:          increase the penalty factor with $\kappa^{(t+1)} = c\kappa^{(t)}$.
7:       **end if**
8:    **end if**
9:    Update $t = t + 1$.
10: **until** converges or $t > T_{\max}$.
11: Obtain $\mathbf{w}_m^*$ according to (2.72).

Remark 2.3 For TPC minimization, the iterative algorithm is generated by finding the minimum point of the penalty function. Since the value of the penalty term is forced to gradually decrease as the penalty factor increases, it is plausible to set a big initial value of penalty factor for reducing the iterations. However, a too large penalty factor makes it difficult to solve the transformed problem or even lead to an unsolvable problem, while a too small penalty factor results in more iterations. Hence, it is important to select a suitable penalty factor for Algorithm 2.4 [40].

2.3.2.3 Optimality Analysis of the Proposed Algorithms

In this section, we focus on the optimality analysis of the two proposed robust design algorithms. For the WSR maximization design, we resort to some approximation methods, i.e., Cauchy-Schwartz inequality, Taylor series expansion and SDR technique, to transform the original problem into a convex problem. Thus, the obtained solutions might be not optimal. But, the approximations will not shrink the feasible region too much over the original problem in the case that the value of uncertainty region $\varepsilon_{m,n}$ is small, the step that updating the Taylor expansion point to approach the constraint (2.51g) in the iterations, and the fact that SDR technique is tight. Hence, it is easy to be verified that the solution we obtain during the iterations ensures the satisfaction of its original problem. For the TCP minimization design, we adopt the SDR technique to overcome the non-convexity issues in the original problem. To guarantee the rank-one constraint, an iterative function based on a penalty function is proposed. Specifically, a new objective function is constructed with the penalty function according to the rank-one constraint. The optimal solution of the primal constrained problem is gradually approached by finding the minimum point of the new objective function through a series of penalty factors. In fact, the value of the penalty term is forced to gradually decrease as the penalty factor increases. Thus, when the penalty factors go to infinity, the minimum point of the new objective function is the optimal solution of the original problem.

2.3.2.4 Complexity Analysis of the Proposed Algorithms

In this section, we analyze the computational complexity of the two proposed robust design algorithms. It is obvious that problems (2.54) and (2.70) are standard SDP problems, only involving LMI, second-order cone (SOC) and affine constraints, which can be effectively solved by using the interior-point method (IPM). According to the computational complexity analysis of a generic IPM for solving the SDP problem in the *Lecture* 6 of [39], for a given $\varepsilon > 0$, the computational cost for an ε-optimal solution is in the order of $\ln (1/\varepsilon) \delta$, where δ is the barrier parameter measuring the geometric complexity of the conic constraints. For simplicity of analysis of barrier parameter δ, we assume that the decision variables n in problems (2.54) and (2.70) are real-valued. First, for the problem (2.54), it has $2M$ LMI constraints of size N_t, $M + 2K$ SOC constraints of size N_t, and $3K$ affine constraints. Second, problem (2.70) has $2K$ LMI constraints of size $N_t + 1$, M LMI constraints of size N_t, and $2K$ affine constraints. Moreover, the number of decision variables n is on the order of $K N_t^2$. Based on the above analysis, the per-iteration complexities for our proposed algorithms are shown in Table 2.2. In addition, to visualize the complexity, we present the computing time for per-iteration of Algorithms 2.3 and 2.4 as the

Table 2.2 Computation complexity analysis of proposed algorithms

Algorithms	Complexity is in order of $\ln(1/\varepsilon)\,\delta$, where $n = \mathcal{O}(KN_t^2)$
Algorithm 2.3	$\delta = \sqrt{(3M+2K)N_t + 3K} \cdot n \cdot$ $\left[(3M+2K)N_t^3 + 3K + n(3M+2K)N_t^2 + 3Kn + n^2\right]$
Algorithm 2.4	$\delta = \sqrt{(M+2K)N_t + 4K} \cdot n \cdot$ $\left[2K(N_t+1)^3 + MN_t^3 + 2K + (2K+M)N_t^2 + 2nKN_t + 4Kn + n^2\right]$

numbers of UEs and BS antennas increase, which is solved by CVX at the SDPT3 solver in the Windows 7 operating system, cf. Fig. 2.12, where $N_m = N = 4, \eta_{m,n} = \eta = 0.05, \varepsilon_{m,n} = \varepsilon = 0.1, q_{m,n} = q_0 = 10\,\text{mW}, \gamma_{m,n} = \gamma_0 = 0.2\,\text{dB}$. It is seen that compared to the exhaustive search, the proposed algorithms have low computational complexity. Hence, they are appealing in the cellular IoT with massive access.

2.3.3 Numerical Results

This section provides some numerical results to validate the robustness and effectiveness of the proposed algorithms for massive SWIPT in the cellular IoT. Unless otherwise stated, we set $N_t = 24, K = 16, M = 4, N_m = N = 4, \eta_{m,n} = \eta = 0.05, \sigma_0^2 = 0.1, \sigma_1^2 = 1, \varepsilon_{m,n} = \varepsilon = 0.1, \alpha_{m,n} = n/\sum_{i=1}^{N_m} i, \rho_{m,n} = 0.8$ and $\theta_{m,n} = 1, \forall m, n$. In addition, we use SNR (in dB) to denote the ratio of transmit power at the BS and the noise variance. For the non-linear EH model, we set $M_{m,n} = 24\,\text{mW}, a_{m,n} = 150$

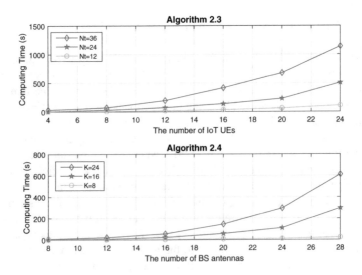

Fig. 2.12 Computing time for per-iteration of Algorithms 2.3 and 2.4

Fig. 2.13 Convergence behavior of the proposed Algorithm 2.3

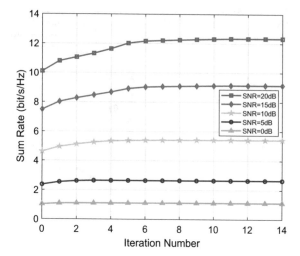

Fig. 2.14 Performance comparison between NOMA scheme and OMA scheme

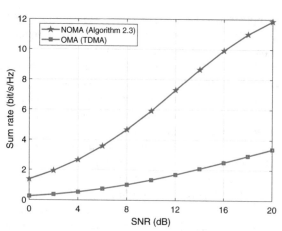

and $b_{m,n} = 0.014$ according to the practical circuit parameters provided by [30]. Without loss of generality, all IoT UEs have the same required EH threshold $q_{m,n} = q_0 = 10\,\text{mW}$ and the same required SINR threshold $\gamma_{m,n} = \gamma_0 = 0.2\,\text{dB}$.

First, we examine the convergence behaviors of the proposed Algorithm 2.3 under different SNR values in Fig. 2.13. It is seen that Algorithm 2.3 has a quick convergence at lower SNR, while it requires a little more number of iterations as SNR increases. In general, Algorithm 2.3 converges no more than 10 iterations on average under different SNR values, which means that the complexity of Algorithm 2.3 is affordable due to a low computational cost of per-iteration as shown in Table I.

Figure 2.14 compares the sum rate performance of the proposed NOMA scheme and the conventional OMA scheme, i.e., TDMA. It is intuitive that the proposed Algorithm 2.3 has a significant performance gain over the TDMA. This is because

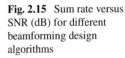

Fig. 2.15 Sum rate versus SNR (dB) for different beamforming design algorithms

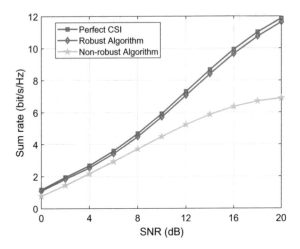

compared to TDMA, the proposed Algorithm 2.3 effectively takes advantage of spatial multiplexing provided by the BS with a multiple-antenna array, and thus can substantially improve the performance.

Then, we illustrate the performance gain of the proposed Algorithm 2.3 over the non-robust algorithm (Zero-forcing beamforming) in Fig. 2.15. Note that the case of perfect CSI means that perfect CSI is available for beamforming design, which can achieve the performance upper bound of massive SWIPT in the cellular IoT. It is seen that Algorithm 2.3 taking into account the norm-bounded channel estimation error is superior to the non-robust one. In particular, the performance gap between Algorithm 2.3 and the non-robust algorithm increases as SNR increases. In addition, the performance loss of Algorithm 2.3 over the perfect CSI case is quite small, which means that Algorithm 2.3 is effective and robust.

Figure 2.16 shows the performance of Algorithm 2.3 with different channel estimation error bounds. Note that the case of $\varepsilon = 0$ means perfect CSI at the BS. It is found that the sum rate decreases as the channel estimation error bound increases. This is because for a given power budget, a large channel estimation error bound requires a higher transmit power to guarantee the worse case of performance and accordingly less power contributes to improving the overall performance. Moreover, the performance loss of Algorithm 2.3 with respect to the perfect CSI case grows with the increase of SNR, which means that ε has a critical impact on the sum rate at high SNR. However, the performance loss due to channel uncertainty is very limited even with $\varepsilon = 0.25$, which reconfirms the robustness of Algorithm 2.3.

Figure 2.17 investigates the capability of Algorithm 2.3 for alleviating the impact of imperfect SIC. It is seen that the performance loss due to imperfect SIC is slight at lower SNR, which means Algorithm 2.3 has a strong capability for alleviating the impact of imperfect SIC in the low SNR region. However, with the increase of SNR, the performance gap with respect to the perfect SIC case ($\eta = 0$) becomes large. This is because in the case of perfect SIC, the intra-cluster interference can be eliminated

Fig. 2.16 Sum rate versus channel estimation error bound ε for different SNR (dB)

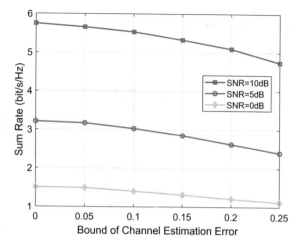

Fig. 2.17 Effects of imperfect SIC and partial CSI on the performance of Algorithm 2.3

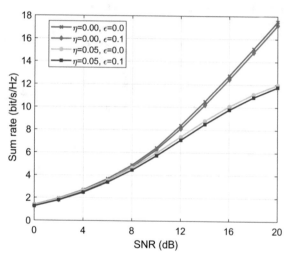

and the sum rate increases as SNR increases, while the sum rate with imperfect SIC ($\eta = 0.05$) will be quickly saturated in the high SNR region. Moreover, it is found that the performance of Algorithm 2.3 is quite close to that of the perfect CSI case. Thus, Algorithm 2.3 can address the practical issues in the cellular IoT.

In Fig. 2.18, we examine the impact of the number of BS antennas N_t on the sum rate of Algorithm 2.3. As expected, the sum rate improves as the number of BS antennas increases, because a large number of BS antennas can provide more array gains to improve the performance. Moreover, since the system is interference limited at high SNR, the sum rate is asymptotically saturated as the SNR increases. Therefore, it is possible to further improve the performance at high SNR by increasing the number of BS antennas.

Fig. 2.18 Impact of the number of BS antennas N_t on the performance of Algorithm 2.3

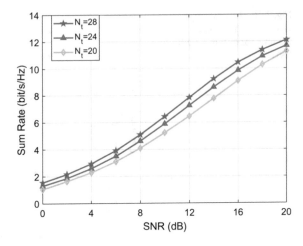

Fig. 2.19 Convergence behavior of the proposed Algorithm 2.4

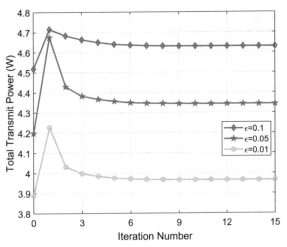

Next, we show the convergence behaviors of Algorithm 2.4 under different channel estimation error bounds in Fig. 2.19. Intuitively, after the first iteration, system finds a better point as its initial point. Similar to Algorithms 2.3, and 2.4 under different conditions converges after no more than 10 times iterations as long as we set a feasible penalty factor. Moreover, it is found that a large channel estimation error bound always consumes more transmit power, which does correspond with the result in Fig. 2.16.

In Fig. 2.20, we reveal the relationship between the TPC and the required minimum SINR γ_0 for different channel estimation error bounds ε. It is observed that extra transmit power is consumed when channel estimation error is involved due to the performance compensation. Further, the total transmit power is sensitive to the channel estimation error bound and increases along with it especially at high SINR

Fig. 2.20 Impact of channel estimation error bound ε on the performance of Algorithm 2.4

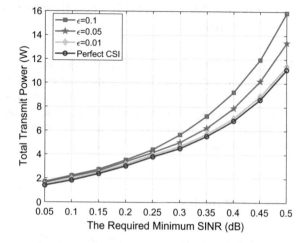

Fig. 2.21 Influence of imperfect SIC and EH threshold on the performance of Algorithm 2.4

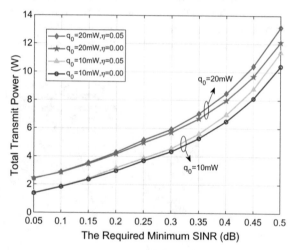

threshold. For instance, all curves almost overlap when the SINR threshold γ_0 is below 0.2 dB, but the gaps between curves grow larger as the minimum required SINR increases.

Finally, we check the impact of imperfect SIC on the power consumption of Algorithm 2.4 at different EH thresholds with a channel estimation error bound $\varepsilon = 0.05$. From Fig. 2.21, it is seen that Algorithm 2.4 with the EH threshold $q_0 = 20$ mW consumes more transmit power than that with the EH threshold $q_0 = 10$ mW. In addition, the case of imperfect SIC ($\eta = 0.05$) always requires more transmit power. This is because there exists a high residual intra-cluster interference after SIC in practical systems. Moreover, the gap between perfect SIC ($\eta = 0.00$) and imperfect SIC is quite small especially at low SINR threshold, which reconfirms that Algorithm 2.4 has a strong capability of alleviating the impact of imperfect SIC.

2.4   Outage-Constrained Robust Design with Channel Estimation Error

In addition to the quantization, channel estimation is another source of CSI errors, which is mainly affected by noise and co-channel interference. In general, the channel estimation errors are modeled by a Gaussian stochastic process. In this section, we provide a comprehensive design framework, including CSI acquisition, signal construction, ID, and EH. According to the characteristic of channel estimation errors, the outage-constrained robust design is proposed.

2.4.1   System Model

In this section, we consider a sustainable B5G cellular IoT network operated in time division duplex (TDD) mode, where a BS equipped with N_t antennas communicates with K single-antenna IoT UEs, cf. Fig. 2.22. It is worth pointing out that in the B5G cellular IoT, the number of BS antennas N_t is usually very large, e.g., $N_t \geq 64$

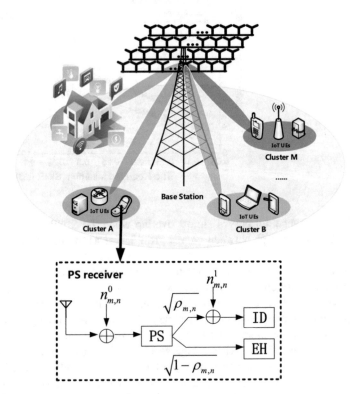

Fig. 2.22 A model of a sustainable B5G cellular IoT network

[41, 42]. For facilitating sustainable communications, UEs divide the received radio frequency (RF) signal into two streams, one for ID and the other for EH. Hence, the massive IoT UEs can survive without batteries.

To realize low-complexity sustainable communications of a massive number of IoT UEs over limited radio spectrum, the BS adopts a multiple-antenna NOMA technique via combining user clustering and spatial beamforming by exploiting its large spatial degrees of freedom. First, as shown in Fig. 2.22, IoT UEs in the identicalbreak spatial direction[4] but with distinct propagation distances are partitioned into a cluster. Without loss of generality, we assume that the K UEs are arranged into M clusters, and the mth cluster contains N_m UEs. For convenience, we use $UE_{m,n}$ to denote the nth UE in the mth cluster in the following. Then, spatial beamforming is conducted to improve the performance of sustainable communications. To design the spatial beamforming scheme, the BS has to know CSI. Herein, we provide a low-cost CSI acquisition method according to the characteristic of user clustering for the B5G cellular IoT with massive access. Specifically, at the beginning of each time slot, the UEs in the mth ($\forall m \in [1, M]$) cluster send a common pilot sequence $\Phi_m \in \mathbb{C}^{1\times\tau}$ with τ symbols over the uplink channels. In spite of the same pilots in identical clusters, the pilots across clusters are pairwise orthogonal, namely $\Phi_j\Phi_i^H = 0$ and $\Phi_j\Phi_j^H = 1, \forall i \neq j$. Thus, the length of the pilot sequence τ can be shortened significantly, especially in the context of massive access. Based on the non-orthogonal pilot sequences, the received signal at the BS is given by

$$\mathbf{Y} = \sum_{j=1}^{M}\sum_{i=1}^{N_j}\sqrt{\delta_{j,i}Q_{j,i}\tau}\mathbf{h}_{j,i}\Phi_j + \mathbf{N}, \qquad (2.73)$$

where $\delta_{j,i}$ and $\mathbf{h}_{j,i}$ denote path loss and small-scale fading vector for the channel from the BS to the $UE_{j,i}$, respectively. It is reasonably assumed that $\delta_{j,i}$ maintains constant during a relatively long time, and is known at the BS perfectly by the long-term statistics. $\mathbf{h}_{j,i}$ is an independently and identical distribution (i.i.d.) zero mean and unit variance complex Gaussian distribution random variable, which keeps unchanged during a time slot, but independently fades over time slots. Moreover, $Q_{j,i}$ and $\mathbf{N}$ are the transmit power of pilot sequence at the $UE_{j,i}$ and an AWGN matrix with zero mean and unit variance at the BS, respectively. Since the UEs in a cluster utilize the same pilot sequence, we only need to estimate the effective CSI for each cluster. Now, we give an example of estimating the effective CSI $\mathbf{h}_m$ for the mth cluster. First of all, right-multiplying $\mathbf{Y}$ by Φ_m^H, we have

---

[4]By applying all kinds of means/technologies, e.g., GPS and auto-tracking system, spatial directions of UEs can be obtained.

$$\mathbf{Y}\boldsymbol{\Phi}_m^H = \sum_{i=1}^{N_j} \sqrt{\delta_{j,i} Q_{j,i} \tau} \mathbf{h}_{m,i} \boldsymbol{\Phi}_m \boldsymbol{\Phi}_m^H + \mathbf{N}\boldsymbol{\Phi}_m^H$$

$$= \sqrt{\sum_{i=1}^{N_j} \delta_{j,i} Q_{j,i} \tau} \mathbf{h}_m + \mathbf{N}\boldsymbol{\Phi}_m^H, \qquad (2.74)$$

where $\mathbf{h}_m = \dfrac{\sum_{n=1}^{N_m} \sqrt{\delta_{m,n} Q_{m,n} \tau} \mathbf{h}_{m,n}}{\sqrt{\sum_{i=1}^{N_m} \delta_{m,i} Q_{m,i} \tau}}$. Then, by using the MMSE channel estimation method, the estimated CSI $\hat{\mathbf{h}}_m$ is given by

$$\hat{\mathbf{h}}_m = \frac{\sqrt{\sum_{n=1}^{N_m} \delta_{m,n} Q_{m,n} \tau}}{1 + \sum_{i=1}^{N_m} \delta_{m,i} Q_{m,i} \tau} \left(\sum_{i=1}^{N_m} \sqrt{\delta_{m,i} Q_{m,i} \tau} \mathbf{h}_{m,i} + \left(\boldsymbol{\Phi}_m \otimes \mathbf{I}_{N_t} \right) \mathrm{vec}(\mathbf{N}) \right). \quad (2.75)$$

Based on (2.75), the relationship between the real CSI $\mathbf{h}_{m,n}$ for $\mathrm{UE}_{m,n}$ and the estimated CSI $\hat{\mathbf{h}}_m$ for the mth cluster can be written as

$$\mathbf{h}_{m,n} = \sqrt{\chi_{m,n}} \hat{\mathbf{h}}_m + \mathbf{e}_{m,n}, \qquad (2.76)$$

where $\mathbf{e}_{m,n}$ is the channel estimation error, which is independent of $\hat{\mathbf{h}}_m$. According to the property of the MMSE channel estimation method, $\mathbf{e}_{m,n}$ is distributed as $\mathscr{CN}\left(\mathbf{0}, \mathbf{C}_{m,n}\right)$ with $\mathbf{C}_{m,n} = \left(1 - \chi_{m,n}\right)\mathbf{I}$ being the associated channel error covariance matrix and $\chi_{m,n} = \frac{\delta_{m,n} Q_{m,n} \tau}{1 + \sum_{i=1}^{N_m} \delta_{m,i} Q_{m,i} \tau} > 0$ denoting the correlation coefficient between $\mathbf{h}_{m,n}$ and $\hat{\mathbf{h}}_m$, namely the CSI accuracy. Due to the sharing of a pilot sequence, the CSI accuracies among the UEs in a cluster are coupled. For instance, adding up all CSI accuracies in the mth cluster, we obtain

$$\sum_{n=1}^{N_m} \chi_{m,n} = \sum_{n=1}^{N_m} \frac{\delta_{m,n} Q_{m,n} \tau}{1 + \sum_{i=1}^{N_m} \delta_{m,i} Q_{m,i} \tau} = \frac{Q_m^{\mathrm{total}}}{1 + Q_m^{\mathrm{total}}} = 1 - \frac{1}{1 + Q_m^{\mathrm{total}}} \leq 1, \quad (2.77)$$

where $Q_m^{\mathrm{total}} = \sum_{i=1}^{N_m} \delta_{m,i} Q_{m,i} \tau$ is the total power of the received pilots associated with the mth cluster. Note that since the sum of the CSI accuracies is upper bounded by 1, it is possible to coordinate the CSI accuracies according to the performance requirements of the IoT UEs via controlling transmit powers in a cluster.

Based on the obtained CSI, the BS constructs and broadcasts a superposition coded signal over the downlink channels as below:

$$\mathbf{x} = \sum_{j=1}^{M} \sum_{i=1}^{N_j} \mathbf{w}_j \sqrt{\alpha_{j,i}} s_{j,i}, \tag{2.78}$$

where $s_{j,i}$ is the Gaussian distributed signal of unit norm for the UE$_{j,i}$, $\alpha_{j,i}$ is the intra-cluster power allocation factor with a constraint $\sum_{i=1}^{N_j} \alpha_{j,i} \leq 1$ for coordinating the intra-cluster interference, and $\mathbf{w}_j$ is an N_t-dimensional spatial beam devised for the jth cluster based on estimated CSI for coordinating the inter-cluster interference. Therefore, the received signal at the UE$_{m,n}$ is given by

$$y_{m,n} = \delta_{m,n}^{1/2} \mathbf{h}_{m,n}^{H} \mathbf{x} + n_{m,n}^{0}, \tag{2.79}$$

where $n_{m,n}^{0}$ is AWGN at the UE$_{m,n}$ with variance σ_0^2. For sustainable communications, the UE$_{m,n}$ divides the received signal into ID and EH streams with a $\rho_{m,n} : 1 - \rho_{m,n}$ proportion, via a power splitter, where $0 \leq \rho_{m,n} \leq 1$ is the PS ratio. We first consider the ID stream. In order to improve the signal quality for ID, SIC is conducted if the IoT UE has enough capability. Without loss of generality, we assume that channel gains in the mth cluster are sorted in a descending order[5]

$$\|\delta_{m,1}^{1/2} \mathbf{h}_{m,1}\|^2 \geq \cdots \geq \|\delta_{m,n}^{1/2} \mathbf{h}_{m,N_m}\|^2. \tag{2.80}$$

Based on the principle of SIC in NOMA [4], the ith UE first decodes and removes the interfering signals associated with the N_mth to $(i + 1)$th UEs according to the descending order in (2.80), and then demodulates its desired signal. Due to numerous factors in practice, e.g., the inferior signal quality and hardware limitations of IoT UEs, there may exist decoding errors of the weak interference signal, resulting in residual interference from the weak UEs after SIC, namely imperfect SIC [43, 44]. Under a linear model about imperfect SIC [17, 27], the post-SIC signal at the UE$_{m,n}$ for ID is given by

$$y_{m,n}^{ID} = \underbrace{\sqrt{\rho_{m,n}\alpha_{m,n}}\delta_{m,n}^{1/2}\mathbf{h}_{m,n}^{H}\mathbf{w}_m s_{m,n}}_{\text{desired signal}} + \underbrace{\sqrt{\rho_{m,n}}\delta_{m,n}^{1/2}\mathbf{h}_{m,n}^{H}\mathbf{w}_m \sum_{i=1}^{n-1}\sqrt{\alpha_{m,i}}s_{m,i}}_{\text{intra-cluster interference from stronger UEs}}$$

$$+ \underbrace{\sqrt{\rho_{m,n}}\eta_{m,n}\delta_{m,n}^{1/2}\mathbf{h}_{m,n}^{H}\mathbf{w}_m \sum_{i=n+1}^{N_m}\sqrt{\alpha_{m,i}}s_{m,i}}_{\text{residual intra-cluster interference from weaker UEs after SIC}} + \underbrace{\sqrt{\rho_{m,n}}\delta_{m,n}^{1/2}\mathbf{h}_{m,n}^{H} \sum_{j=1,j\neq m}^{M}\mathbf{w}_j \sum_{i=1}^{N_j}\sqrt{\alpha_{j,i}}s_{j,i}}_{\text{inter-cluster interference}}$$

$$+ \underbrace{\sqrt{\rho_{m,n}}n_{m,n}^{0} + n_{m,n}^{1}}_{\text{noise}}, \tag{2.81}$$

---

[5]To determine the order of SIC, the BS only needs channel gains rather than CSI. In general, each UE knows its own channel gain for coherent signal detection. In practical systems, UEs send channel quality information, namely channel gain, to the BS over the uplink channels. Since channel gain is a real value, the amount of feedback is very small. Hence, the BS can sort channel gains in a cluster, and informs the UEs over the downlink channels.

where the second term of (2.81) is the intra-cluster interference from stronger UEs in a cluster, and the third one is the residual intra-cluster interference from weaker UEs due to imperfect SIC. Note that $\eta_{m,n}$ denotes the imperfect SIC coefficient at the UE$_{m,n}$ acquired by long-term measurement, which ranges from 0 to 1. Specifically, $\eta_{m,n} = 0$ means perfect SIC, $\eta_{m,n} = 1$ denotes no SIC, and $0 < \eta_{m,n} < 1$ refers to imperfect SIC. $n^1_{m,n}$ is the baseband AWGN with variance σ^2_1 caused by the signal conversion from RF band to baseband. Consequently, the received SINR at the ID receiver of UE$_{m,n}$ can be expressed as

$$\Gamma_{m,n} = \frac{\alpha_{m,n}\left|\mathbf{h}^H_{m,n}\mathbf{w}_m\right|^2}{\left|\mathbf{h}^H_{m,n}\mathbf{w}_m\right|^2 \sum\limits_{i=1}^{n-1} \alpha_{m,i} + \eta_{m,n}\left|\mathbf{h}^H_{m,n}\mathbf{w}_m\right|^2 \sum\limits_{i=n+1}^{N_m} \alpha_{m,i} + \sum\limits_{j=1,j\neq m}^{M} \left|\mathbf{h}^H_{m,n}\mathbf{w}_j\right|^2 + \frac{\sigma^2_0}{\delta_{m,n}} + \frac{\sigma^2_1}{\delta_{m,n}\rho_{m,n}}}. \tag{2.82}$$

Then, we consider the EH stream of UE$_{m,n}$, which is given by

$$y^{\text{EH}}_{m,n} = \sqrt{1 - \rho_{m,n}}\left(\delta^{1/2}_{m,n}\mathbf{h}^H_{m,n}\mathbf{x} + n^0_{m,n}\right). \tag{2.83}$$

By using a practical non-linear EH circuit [30], the harvested power at the UE$_{m,n}$ can be expressed as

$$P^{\text{EH}}_{m,n} = \frac{\frac{M_{m,n}}{1+\exp(-a_{m,n}(P^{\text{in}}_{m,n}-b_{m,n}))} - \frac{M_{m,n}}{1+\exp(a_{m,n}b_{m,n})}}{1 - \frac{1}{1+\exp(a_{m,n}b_{m,n})}}, \tag{2.84}$$

where $P^{\text{in}}_{m,n}$ is the input power, which is given by

$$P^{\text{in}}_{m,n} = \left(1 - \rho_{m,n}\right)\left(\sum_{j=1}^{M} \delta_{m,n}\left|\mathbf{h}^H_{m,n}\mathbf{w}_j\right|^2 + \sigma^2_0\right). \tag{2.85}$$

Parameters $M_{m,n}$, $a_{m,n}$ and $b_{m,n}$ are all constants related to the detailed system specifications. Specifically, $M_{m,n}$ accounts for the maximum harvested power at the UE$_{m,n}$ as the EH circuit is saturated, while $a_{m,n}$ and $b_{m,n}$ represent the physical hardware phenomena of the EH circuit, e.g., threshold voltage and leakage current.

In the scenario of sustainable B5G cellular IoT with massive access, interference is an inevitable factor that significantly influences the performance of both ID and EH. As seen from (2.82) to (2.84), the interference reduces the quality of the received signal for ID, but increases the amount of the received signal for EH. Hence, it makes sense to coordinate the interference for improving the overall performance.

2.4.2 Problem Formulation and Optimization Design

In this section, we aim to design and optimize the sustainable communication scheme for B5G cellular IoT with massive access. Since the BS only has partial CSI through non-orthogonal channel estimation, it is necessary to design a robust scheme under practical but adverse conditions, e.g., imperfect SIC and non-linear EH.

For significantly improving the overall performance, we focus on jointly optimizing spatial beams at the BS, transmit powers within each cluster and PS ratios at the UEs by maximizing the WSR of all UEs while guaranteeing their EH requirements. According to the statistical characteristic of CSI errors based on the proposed channel estimation method, we attempt to design an outage-constrained robust scheme. In particular, the design of an outage-constrained robust scheme is equivalent to the following optimization problem:

$$\max_{\mathbf{w},\rho,\alpha} \sum_{m=1}^{M} \sum_{n=1}^{N_m} \theta_{m,n} R_{m,n} \tag{2.86a}$$

$$\text{s.t.} \sum_{m=1}^{M} \|\mathbf{w}_m\|^2 \leq P_{\max}, \tag{2.86b}$$

$$\sum_{n=1}^{N_m} \alpha_{m,n} \leq 1, \forall m, \tag{2.86c}$$

$$\alpha_{m,n} \leq \alpha_{m,n+1}, n \in [1, N_m - 1], \tag{2.86d}$$

$$0 \leq \rho_{m,n} \leq 1, \forall m, n, \tag{2.86e}$$

$$\Pr\left\{P_{m,n}^{\text{EH}} \geq q_{m,n}\right\} \geq 1 - p_{m,n}^{\text{out}}, \forall m, n, \tag{2.86f}$$

where $\theta_{m,n} > 0$ denotes the priority for $\text{UE}_{m,n}$, $R_{m,n} = \log_2(1 + \Gamma_{m,n})$ is the achievable rate (in bit/s) of the $\text{UE}_{m,n}$, $P_{\max} > 0$ represents the maximum transmit power at the BS, $q_{m,n} > 0$ is the required minimum power for sustainable communications at the $\text{UE}_{m,n}$, and $p_{m,n}^{\text{out}} \in (0, 1]$ is the tolerant outage probability of EH dependent of the requirement of QoS at the $\text{UE}_{m,n}$. Moreover, $\mathbf{w} = \{\mathbf{w}_1, \ldots, \mathbf{w}_M\}$, $\alpha = \{\alpha_{1,1}, \ldots, \alpha_{M,N_m}\}$ and $\rho = \{\rho_{1,1}, \ldots, \rho_{M,N_m}\}$ are optimization variables sets of spatial beams, intra-cluster power allocation factors and PS ratios, respectively. The constraint (2.86b) and (2.86c) represent the power allocation among the clusters and among the UEs within a cluster, respectively. The constraint (2.86d) is used to facilitate the operation of SIC in a cluster, (2.86e) is a constraint on the PS ratios, and (2.86f) assures that the $\text{UE}_{m,n}$ reaches its required minimum EH value $q_{m,n}$ with a probability $1 - p_{m,n}^{\text{out}}$ at least.

Remark 2.4 Note that according to the performance requirements of IoT UEs, it is possible to achieve a tradeoff between service fidelity and design conservatism. Generally speaking, it is expected to adopt a low EH outage probability $p_{m,n}^{\text{out}}$ for achieving a high service fidelity, while leading to a more conservative design for (2.86). Especially, if we choose a tiny value of the EH outage probability $p_{m,n}^{\text{out}}$ in the design, there may be no feasible solution to (2.86) or even unacceptable performance. In what follows, we assume that there always exist feasible solutions by setting appropriate parameters.

In general, the WSR problem (2.86) is non-convex, for which finding its global optimal solution is quite difficult and challenging. For one thing, the objective function is non-convex since all optimization variables are coupled. For another, the EH constraint $P_{m,n}^{\mathrm{EH}} \geq q_{m,n}$ is also non-convex with respect to spatial beam $\mathbf{w}_m$ and PS ratio $\rho_{m,n}$, and its probability function has no tractable expression for the considered CSI error model. To solve these challenges, we need to resort to some approximation methods. First, we deal with the probabilistic EH constraint (2.86f). To facilitate processing, we define $E_{m,n} = b_{m,n} - \frac{1}{a_{m,n}} \ln \frac{e^{a_{m,n} b_{m,n}} \left(M_{m,n} - q_{m,n} \right)}{q_{m,n} e^{a_{m,n} b_{m,n}} + M_{m,n}}$. Combining with the result of non-orthogonal channel estimation in (2.76), constraint (2.86f) can be rewritten as

$$
\Pr \left\{ \sum_{j=1}^{M} \delta_{m,n} \left| \left(\sqrt{\chi_{m,n}} \hat{\mathbf{h}}_m + \mathbf{e}_{m,n} \right)^H \mathbf{w}_j \right|^2 + \sigma_0^2 \geq \frac{E_{m,n}}{1 - \rho_{m,n}} \right\} \geq 1 - p_{m,n}^{\mathrm{out}}. \quad (2.87)
$$

Moreover, based on $\mathbf{e}_{m,n} \sim \mathscr{CN}\left(0, \mathbf{C}_{m,n}\right)$, we make

$$
\mathbf{z}_{m,n} = \mathbf{C}_{m,n}^{-1/2} \mathbf{e}_{m,n} \sim \mathscr{CN}\left(0, \mathbf{I}\right). \quad (2.88)
$$

Then, let $\mathbf{T} = \sum_{j=1}^{M} \mathbf{w}_j \mathbf{w}_j^H$, (2.87) can be further transformed as

$$
\Pr \left\{ \mathbf{z}_{m,n}^H \mathbf{C}_{m,n}^{1/2} \mathbf{T} \mathbf{C}_{m,n}^{1/2} \mathbf{z}_{m,n} + 2 \operatorname{Re} \left\{ \sqrt{\chi_{m,n}} \mathbf{z}_{m,n}^H \mathbf{C}_{m,n}^{1/2} \mathbf{T} \hat{\mathbf{h}}_m \right\} \geq \xi_{m,n} \right\} \geq 1 - p_{m,n}^{\mathrm{out}},
$$
$$
\quad (2.89)
$$

where $\xi_{m,n} = \frac{E_{m,n}}{\delta_{m,n}(1 - \rho_{m,n})} - \frac{\sigma_0^2}{\delta_{m,n}} - \chi_{m,n} \hat{\mathbf{h}}_{m,n}^H \mathbf{T} \hat{\mathbf{h}}_m$. Next, we develop a conservative approximation for (2.86f) based on a relaxation-restriction (RAR) methodology [45], which has two steps, i.e., SDR and *Bernstein-type* inequality (BTI) restriction, and can bound the probability of quadratic forms of Gaussian variable involving matrices. Thus, the following lemma is required to transform the probabilistic constraints into the deterministic ones.

Lemma 2.5 (BTI, [45]) *For any complex Hermitian matrix $\mathbf{A} \in \mathbb{C}^{N \times N}$, vector $\mathbf{b} \in \mathbb{C}^{N \times 1}$, scalar $c \in \mathbb{R}$ and outage probability p, there exists vector $\mathbf{x} \sim \mathscr{CN}\left(0, \mathbf{I}_N\right)$ making the outage constraint*

$$
\Pr \left\{ \mathbf{x}^H \mathbf{A} \mathbf{x} + 2 \operatorname{Re} \left\{ \mathbf{x}^H \mathbf{b} \right\} + c \geq 0 \right\} \geq 1 - p \quad (2.90)
$$

holds:

$$\begin{cases} \mathrm{tr}\,(\mathbf{A}) - \sqrt{-2\ln{(p)}}s + \ln{(p)}\,t + c \geq 0, \\ \left\| \begin{bmatrix} \mathrm{vec}\,(\mathbf{A}) \\ \sqrt{2}\mathbf{b} \end{bmatrix} \right\| \leq s, \\ t\mathbf{I}_N + \mathbf{A} \succeq \mathbf{0}, \end{cases} \quad (2.91)$$

where $s > 0$ and $t > 0$ are slack variables.

However, due to the limitation of the objective function in spatial beams, the step of SDR in the RAR methodology has to be discarded. Hence, we adopt a Taylor series expansion $f\,(x) = \sum_{n=0}^{\infty} \frac{f^{(n)}}{n!}(x - x_0)^n + R_n\,(x)$ to approximatively transform the quadratic function $\mathbf{T} = \sum_{j=1}^{M} \mathbf{w}_j \mathbf{w}_j^H$ into a linear one. The first-order Taylor series expansion of $\mathbf{T}$ at $\widetilde{\mathbf{w}}_j$ is given by

$$\widetilde{\mathbf{T}} = \sum_{j=1}^{M} \left(2\,\mathrm{Re}\left\{ \mathbf{w}_j \widetilde{\mathbf{w}}_j^H \right\} - \widetilde{\mathbf{w}}_j \widetilde{\mathbf{w}}_j^H \right), \quad (2.92)$$

where $\widetilde{\mathbf{w}}_j$ is a feasible point of problem (2.86). In order to achieve a better suboptimal solution, we set a feasible initial value at the beginning and then utilize the solution of the last iteration as the feasible point to problem (2.86) at the current round of iterations. By exploiting the BTI in *Lemma 5*, the EH constraint (2.89) can be transformed as

$$\mathrm{tr}\,(\mathbf{B}_{m,n}) - \sqrt{-2\ln{\left(p_{m,n}^{\mathrm{out}}\right)}}r_{m,n} + \ln{\left(p_{m,n}^{\mathrm{out}}\right)}t_{m,n} \geq \widetilde{\xi}_{m,n}, \quad (2.93\mathrm{a})$$

$$\left\| \begin{bmatrix} \mathrm{vec}\,(\mathbf{B}_{m,n}) \\ \sqrt{2}\mathbf{u}_{m,n} \end{bmatrix} \right\| \leq r_{m,n}, \quad (2.93\mathrm{b})$$

$$t_{m,n}\mathbf{I}_N + \mathbf{B}_{m,n} \succeq \mathbf{0}, \quad (2.93\mathrm{c})$$

where $\mathbf{B}_{m,n} = \mathbf{C}_{m,n}^{1/2}\widetilde{\mathbf{T}}\mathbf{C}_{m,n}^{1/2}$, $\mathbf{u}_{m,n} = \sqrt{\chi_{m,n}}\mathbf{C}_{m,n}^{1/2}\widetilde{\mathbf{T}}\hat{\mathbf{h}}_m$, and $\widetilde{\xi}_{m,n} = \frac{E_{m,n}}{\delta_{m,n}(1-\rho_{m,n})} - \frac{\sigma_0^2}{\delta_{m,n}} - \chi_{m,n}\hat{\mathbf{h}}_{m,n}^H\widetilde{\mathbf{T}}\hat{\mathbf{h}}_m$. $r_{m,n} \geq 0$ and $t_{m,n} \geq 0$ are slack variables. It is found that although the transformed EH constraint (2.93) has been convex, the objective function (2.86a) is still non-convex due to coupled multiple variables. Note that in the case of full CSI at the BS, the received SINR $\gamma_{m,n}$ and the MMSE $\varepsilon_{m,n}$ between the transmit and receive signals have the equivalent relationship as $1 + \gamma_{m,n} = \varepsilon_{m,n}^{-1}$ ([10], Lemma 1). In other words, maximizing the achievable rate at $\mathrm{UE}_{m,n}$ is equivalent to minimizing $\log_2\left(\varepsilon_{m,n}\right)$. However, the BS can only obtain partial CSI in the practical B5G cellular IoT, which further increases the difficulty in dealing with the objective function. To tackle this issue, we propose the following theorem.

Theorem 2.2 *According to the relationship between average rate and mean squared error (MSE), the original objective function (2.86a) can be approximated as*

$$\min_{\mathbf{w}, \mathbf{v}, \alpha, \rho} \sum_{m=1}^{M} \sum_{n=1}^{N_m} \theta_{m,n}\log_2\left(\overline{\mathrm{MSE}}_{m,n}\right), \quad (2.94)$$

where $\overline{\text{MSE}}_{m,n}$ *denotes the average MSE* $E\{\text{MSE}_{m,n}\}$ *for* $UE_{m,n}$ *and* $\mathbf{v} = \{v_{1,1}, \ldots, v_{M,N_m}\}$ *is a set of the ID receivers of UEs.*

Proof Please refer to Appendix D. □

It is found that the sum of logarithmic function in (2.94) still hinders us to further solve this problem. To this end, we introduce a weight variable $\beta_{m,n}$ for the ID receiver at the $UE_{m,n}$ to make the problem more tractable [33, 47], and thus the logarithmic function can be replaced with the following term:

$$\min_{\mathbf{w},\mathbf{v},\alpha,\rho,\beta} \sum_{m=1}^{M} \sum_{n=1}^{N_m} \theta_{m,n} \left(\beta_{m,n} \overline{\text{MSE}}_{m,n} - \log_2 \left(\beta_{m,n} \right) \right), \qquad (2.95)$$

where $\beta = \{\beta_{1,1}, \ldots, \beta_{M,N_m}\}$ is a set of the weight variables. Note that the objective function (2.95) achieves its minimum value if and only if $\beta_{m,n}$ is equal to $\overline{\text{MSE}}_{m,n}$ with the minimum average MSE (MAMSE) receiver. Hence, the original problem (2.86) is changed as

$$\min_{\mathbf{w},\mathbf{v},\mathbf{r},\mathbf{t},\alpha,\rho,\beta} \sum_{m=1}^{M} \sum_{n=1}^{N_m} \theta_{m,n} \left(\beta_{m,n} \overline{\text{MSE}}_{m,n} - \log_2 \left(\beta_{m,n} \right) \right)$$

$$\text{s.t.}\quad (2.86\text{b})-(2.86\text{e}),\ (2.93\text{a})-(2.93\text{c}),$$

$$r_{m,n} \geq 0,\ t_{m,n} \geq 0. \qquad (2.96)$$

Based on the expression of $\overline{\text{MSE}}_{m,n}$ in (2.116), it is easy to know that problem (2.96) is not a joint convex problem of $\mathbf{w}, \mathbf{v}, \mathbf{r}, \mathbf{t}, \alpha, \rho, \beta$, but it is convex for each optimization variable $\mathbf{w}, \mathbf{v}, \alpha, \rho, \beta$ (where $\mathbf{r}, \mathbf{t}$ are slacked variables for spatial beams and PS ratios). Thereby, the block coordinate descent method can be applied to solve this problem. In other words, we sequentially optimize one variable by fixing the others until they approach a stationary point in the iterations. More specifically, first, we set the weight variable $\beta_{m,n}$ equal to $\overline{\text{MSE}}_{m,n}^{-1}$. Second, there is a closed-form solution $v_{m,n} = \sqrt{\rho_{m,n} \alpha_{m,n} \delta_{m,n} \chi_{m,n}} \mathbf{w}_m^H \hat{\mathbf{h}}_m O_{m,n}^{-1}$, which is derived in Appendix A, for the MAMSE receiver to maximize the lower bound on the average WSR. Finally, since spatial beam $\mathbf{w}_m$, intra-cluster power allocation factor $\alpha_{m,n}$ and PS ratio $\rho_{m,n}$ only involved with multiple convex constraints, they can be iteratively optimized by off-the-shelf tools, e.g., CVX [36]. In summary, the outage-constrained robust design for sustainable B5G cellular IoT can be described as Algorithm 2.5.

Remark 2.5 Since the proposed algorithm is iterative in nature, the initialization is important to achieve quick convergence, and affects the final performance. For the first-order Taylor series expansion of $\mathbf{T}$ at point $\tilde{\mathbf{w}}_m$, $\forall m$ in constraint (2.92), the closer the expansion point $\tilde{\mathbf{w}}_m$ is to the variable $\mathbf{w}_m$, the smaller the approximation error is. To get a better approximation, we set $\tilde{\mathbf{w}}_m^{(t+1)} = \mathbf{w}_m^{(t)}$ in the iterations, and make $\tilde{\mathbf{w}}_m^{(0)} = \sqrt{\frac{P_{\max}}{M}}[1, 0, \cdots, 0]^T$ to satisfy the transmit power constraint. In addition,

$\alpha_{m,n}^{(0)} = \frac{1}{N_m}$ and $\rho_{m,n}^{(0)} = \rho = 0.5, \forall m, n$ are used to satisfy the intra-cluster power allocation constraint and PS ratio constraint in the first iteration, respectively.

Algorithm 2.5 Robust Design for Sustainable B5G Cellular IoT

Input: $N_t, K, M, N_m, q_{m,n}, \eta_{m,n}, \theta_{m,n}, \chi_{m,n}, \sigma_0^2, \sigma_1^2, P_{\max}$.
Output: $\mathbf{w}_m, \alpha_{m,n}, \rho_{m,n}$

1: **Initialize** $\tilde{\mathbf{w}}_m^{(0)} = \sqrt{\frac{P_{\max}}{M}}[1, 0, \cdots, 0]^T, \alpha_{m,n}^{(0)} = \frac{1}{N_m}, \rho_{m,n}^{(0)} = \rho = 0.5, \forall m, n,$ the   WSR
 $R^{(0)} = 0$, convergence accuracy $\Delta = 0$, the maximum number of iterations $T_{\max} = 30$, and
 iteration index $t = 1$.
2: **Set** the auxiliary variable $\beta_{m,n} = \overline{\text{MSE}}_{m,n}^{-1}$, and the MAMSE receiver $v_{m,n} = \sqrt{\rho_{m,n}\alpha_{m,n}\delta_{m,n}\chi_{m,n}}\mathbf{w}_m^H\hat{\mathbf{h}}_m O_{m,n}^{-1}$
3: **while** $\Delta > 0.01$ and $t < T_{\max}$ **do**
4:    Solve problem (2.96) by CVX with fixed $\rho_{m,n}^{(t)}$ and $\alpha_{m,n}^{(t)}$, then obtain $\mathbf{w}_m^{(t)}$ and update $\tilde{\mathbf{w}}_m^{(t+1)} = \mathbf{w}_m^{(t)}$;
5:    Solve problem (2.96) by CVX with fixed $\mathbf{w}_m^{(t)}$ and $\rho_{m,n}^{(t)}$, then obtain $\alpha_{m,n}^{(t)}$;
6:    Solve problem (2.96) by CVX with fixed $\mathbf{w}_m^{(t)}$ and $\alpha_{m,n}^{(t)}$, then obtain $\rho_{m,n}^{(t)}$;
7:    Update $\beta_{m,n}^{(t)}, v_{m,n}^{(t)}$ and compute $R^{(t)}$;
8:    Update $\Delta = R^{(t)} - R^{(t-1)}$;
9:    Update $t = t + 1$;
10: **end while**

Convergence Analysis: Since the problem (2.96) in terms of each optimization variable is convex, the optimal solution to each sub-problem can be obtained using CVX in each iteration. Based on the update steps in Algorithm 2.5, the solutions in the t-th iteration are the feasible solution in the $(t + 1)$-th iteration for problem (2.96), which means that the objective value obtained in the $(t + 1)$-th iteration is less than or equal to that in the t-th iteration. In other words, the lower bound on the average WSR is nondecreasing after each iteration. Furthermore, due to the constraint of transmit power at the BS, the lower bound on the average WSR is bounded. Such a conclusion guarantees the convergence of the proposed algorithm.

Complexity Analysis: It is seen that the primary computational complexity of the proposed Algorithm 2.5 mainly stems from step 4, i.e., optimizing the spatial beams. Since problem (2.96) in terms of optimizing the spatial beams is a SOCP problem due to only LMI and SOC constraints, it can be solved efficiently by the IPM ([45], Section V-A). Specifically, it has $5K$ LMI constraints of size 1, K LMI constraints of size N_t, M SOC constraints of dimension N_t, and K SOC constraints of dimension $N_t + 1$. Thus, for a given $\varepsilon > 0$, the per-iteration complexity for an ε-optimal solution by IPM is in order of $\ln(\varepsilon^{-1})\varpi$, where $\varpi = \sqrt{(2K + M)N_t + 6K}[(2K + M)N_t^3 + (3K + 2Kn + Mn)N_t^2 + 2K(n + 1)(N_t + 3) + K + n^2]n$ and the decision variable $n = \mathcal{O}(KN_t^2)$. Thus, Algorithm 2.5 can obtain a sub-optimal solution in polynomial time [35].

Table 2.3 Simulation parameters

Parameters	Values
Number of BS antennas and IoT UEs	$N_t = 64$, $K = 36$
User clustering	$M = 12$, $N_m = N = 3$
CSI accuracy coefficient and SIC coefficient	$\chi_{m,n} = \chi = 0.33$, $\eta_{m,n} = \eta = 0.05$
Priority of IoT UEs	$\theta_{m,n} = \theta = 1$
Total transmit power budget available at the BS	$P_{\max} = 30$ dBm
Noise powers	$\sigma_0^2 = -70$ dBm, $\sigma_1^2 = -50$ dBm
Non-linear EH model	$M_{m,n} = 24$ mW, $a_{m,n} = 150$, $b_{m,n} = 0.014$ [30]
EH outage probability	$p_{m,n}^{\text{out}} = p_{\text{out}} = 0.1$
Minimum required EH threshold	$q_{m,n} = q_0 = -10$ dBm

2.4.3   Numerical Results

In this section, we provide extensive simulation results to validate the robustness and effectiveness of the proposed Algorithm 2.5 for the sustainable B5G cellular IoT. Unless extra specification, the simulation parameters are set as in Table. 2.3. The path loss model is given by $PL_{dB} = 128.1 + 37.6 \log_{10}(d)$ [48], where d (km) is the distance between the IoT UE and the BS. Without loss of generality, it is assumed that all IoT UEs are randomly distributed within 20 m from the BS.

First, we present the convergence behavior of Algorithm 2.5. As seen in Fig. 2.23, the WSR increases monotonically as expected, and converges to a stationary value within 10 iterations on average under different maximum transmit power budget $P_{\max}$ at the BS. Especially, for a small value of $P_{\max}$, it is almost a straight line. That

Fig. 2.23 Convergence behavior of the proposed Algorithm 2.5

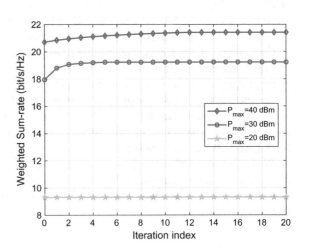

Fig. 2.24 WSR (bit/s/Hz)
versus the maximum
transmit power at the BS
P_{max} (dBm) for different
schemes

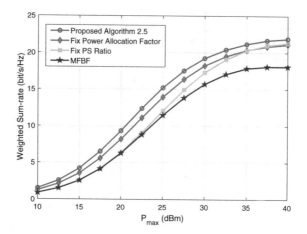

implies Algorithm 2.5 converges quickly and thus has low complexity, which is quite appealing to B5G cellular IoT with massive access.

Second, we show the performance gain of Algorithm 2.5 over several baseline resource allocation algorithms in Fig. 2.24, i.e., a fixed intra-cluster power allocation factors algorithm with $\alpha_{m,n} = n / \sum_{i=1}^{N_m} i$, a fixed PS ratios algorithm with $\rho_{m,n} = \rho = 0.5$ and a match filtering beamforming (MFBF) algorithm with $\mathbf{w}_m = \frac{\hat{\mathbf{h}}_m}{\|\hat{\mathbf{h}}_m\|}$. It is seen that the proposed Algorithm 2.5 substantially outperforms other algorithms in the whole P_{max} region, since it jointly optimizes all variables affecting the performance. Specifically, the gap between Algorithm 2.5 and the fixed power allocation factor algorithm is almost keep unchanged, which means that the intracluster power allocation has a great impact on the performance. Besides, it is found that the gap between Algorithm 2.5 and the fixed PS ratios algorithm decreases with the increment of P_{max}. This is because for a high transmit power, the power consumption for EH is small and the impact of the PS ratio can be ignored. Moreover, although the MFBF algorithm has a lower complexity compared to Algorithm 2.5, it leads to a large performance degradation. Thus, it is a good choice to apply Algorithm 2.5 achieving a balance between system performance and implementation complexity in practical B5G cellular IoT.

Third, we verify the capability of Algorithm 2.5 for mitigating the effect of imperfect SIC in Fig. 2.25. It is seen that the higher imperfect SIC coefficient, the lower WSR. This is because imperfect SIC arises a high residual intra-cluster interference, resulting in severe performance degradation. Moreover, for a low transmit power at the BS, there is a small gap between perfect SIC and imperfect one, which means Algorithm 2.5 can make use of the residual interference effectively. Hence, even in the adverse case of partial CSI at the BS and imperfect SIC at the UEs, Algorithm 2.5 has a promising potential of enhancing the performance.

Next, we check the impact of the number of BS antennas N_t on the performance of Algorithm 2.5 in Fig. 2.26. It is intuitive that the WSR improves as the number of

Fig. 2.25 WSR (bit/s/Hz) versus the maximum transmit power at the BS P_{max} (dBm) for different imperfect SIC coefficient η

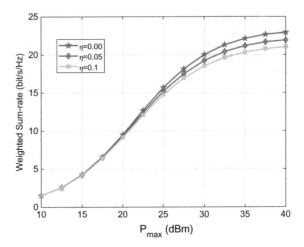

Fig. 2.26 WSR (bit/s/Hz) versus the maximum transmit power at the BS P_{max} (dBm) for different the number of BS antenna N_t

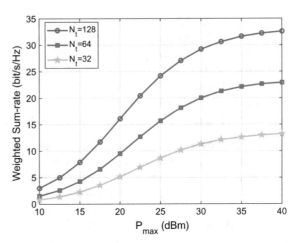

BS antennas increases, since UEs can acquire more array gains for enhancing the performance. However, dominated by interference, the WSR will be saturated as the transmit power is sufficiently large. Thus, it is possible to increase the number of BS antennas for further improving the performance at high transmit power. Moreover, it is worth pointing out that Algorithm 2.5 can achieve a high performance gain even in the case that the BS is equipped with a not so large number of antennas, e.g., $N_t = 32$, which means Algorithm 2.5 has a strong capability of supporting massive access even with a not so large number of BS antennas.

Figure 2.27 reveals the influences of user clustering mode and CSI accuracy on the performance of Algorithm 2.5. As mentioned above, the number of clusters M determines the inter-cluster interference and the complexity of beamforming design, while the number of UEs N in a cluster affects the intra-cluster interference and the complexity of SIC. Moreover, the CSI accuracy in the context of non-orthogonal channel

Fig. 2.27 WSR (bit/s/Hz) versus the maximum transmit power at the BS P_{max} (dBm) for different user clustering modes

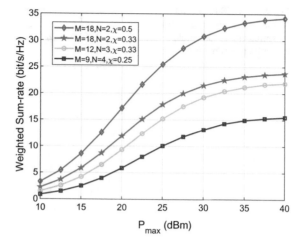

Fig. 2.28 WSR (bit/s/Hz) versus CSI accuracy χ for different algorithms

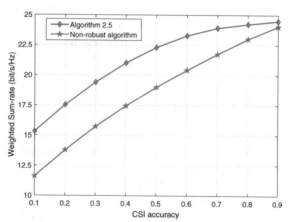

estimation is limited by the number of UEs N in a cluster, i.e., $\sum_{n=1}^{N_m} \chi_{m,n} \leq 1$. Without loss of generality, we assume that average CSI accuracy can achieve the limit, namely $M = 18$, $N = 2$, $\chi = 0.5$ as mode 1, $M = 18$, $N = 2$, $\chi = 0.33$ as mode 2, $M = 12$, $N = 3$, $\chi = 0.33$ as mode 3, and $M = 9$, $N = 4$, $\chi = 0.25$ as mode 4. Obviously, the impact of CSI accuracy distinctly exceeds that of user clustering mode. However, in the context of massive access, a small value of N requires longer pilot sequences for channel estimation, resulting in outdated CSI. Besides, it is hard to support massive access with finite antennas at the BS for a small number of UEs in a cluster. Thus, it is desired to select an appropriate user clustering mode to strike a balance between system performance and computational complexity.

Figure 2.28 illustrates the impact of CSI accuracy, and the performance gain of Algorithm 2.5 over the non-robust algorithm. Note that the non-robust algorithm regards the estimated channel $\hat{\mathbf{h}}_m$, $\forall m$ as the real channel in the design. It is seen that the WSR increases as the CSI accuracy improves for Algorithm 2.5. This is

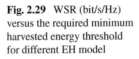

Fig. 2.29 WSR (bit/s/Hz) versus the required minimum harvested energy threshold for different EH model

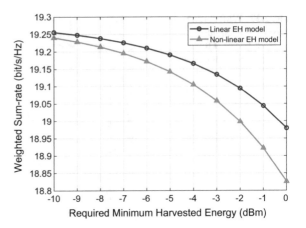

because a high CSI accuracy requires less transmit power to overcome the channel uncertainty and accordingly more power contributes to improving the overall performance. Moreover, Algorithm 2.5 taking into account the CSI uncertainty in the design always outperforms the non-robust one in the whole region, and the gap between them enlarges as the CSI accuracy decreases, which demonstrates the effectiveness and robustness of Algorithm 2.5. In particular, in the case of user clustering, the sum of CSI accuracies in a cluster is upper bounded by 1. For example, if the number of UEs in the mth cluster is equal to 3, i.e., $N_m = 3$, the average CSI accuracy is less than $1/3$. In other words, for the cellular IoT with a massive number of UEs, the CSI accuracy is low for each UE, which further proves the necessity of the proposed algorithm.

Figure 2.29 shows the performance versus EH threshold for linear and non-linear EH models, where we set the energy conversion efficiency $\vartheta_{m,n} = \vartheta = 0.5$ for the linear EH model. It is intuitive that the WSR of Algorithm 2.5 decreases with the growth of required minimum harvested energy for each UE. This is because more transmit power is consumed to meet the increasing EH threshold. Moreover, the system designed with the linear EH model always outperforms than that designed with the non-linear EH model, since the harvesting energy with the non-linear EH model will be saturated but that with the linear EH model grows linearly. Thus, the non-linear EH model is more accurate to capture the characteristic of EH in practice.

Finally, we examine the effect of EH outage probability on the performance of Algorithm 2.5. As can be observed in Fig. 2.30, the WSR grows with the increment of the EH outage probability. This corresponds with the fact that a lower outage probability leads to a more stricter constraint of harvested energy for the system, which will consume more resources to transfer power for the EH while a relevant reduction of power for ID. Moreover, it is found that the system restricted with a high EH threshold has a lower WSR, yet it can achieve a great performance gain as EH outage probability improves. This is because for a given transmit power, it will require more power to meet a high EH requirement of UE, accordingly resulting

Fig. 2.30 WSR (bit/s/Hz) versus the EH outage probability (%) for different EH threshold

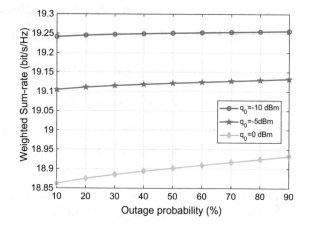

in a large performance loss, and is more sensitive to the EH outage probability. In addition, as analyzed in Remark 2.1, it is important to choose a proper EH outage probability in the design for achieving a balance between service fidelity and design conservatism.

2.5   Conclusion

In this chapter, we exploited the benefits of multiple-antenna NOMA to support massive access for convergence of energy and communication in B5G cellular IoT. In particular, we proposed comprehensive design frameworks and optimization algorithms for both full CSI and imperfect CSI under practical conditions. Extensive simulation results demonstrated that the proposed algorithms were effective to alleviate the impacts of practically adverse factors, e.g., imperfect SIC, non-linear EH, channel uncertainty, and limited radio spectrum, and thus enhanced their applicabilities in B5G cellular IoT.

Appendix

A: Proof of Lemma 2.1

For the received signal for ID $y_{m,n}^{\text{ID}}$, the MSE at the nth information UE in the mth cluster can be given by

$$\text{MSE}_{m,n} = \text{E}\left[\left(v_{m,n} y_{m,n}^{\text{ID}} - s_{m,n}\right)\left(v_{m,n} y_{m,n}^{\text{ID}} - s_{m,n}\right)^{H}\right], \qquad (2.97)$$

where $v_{m,n}$ represents the receiver at the nth information UE in the mth cluster. Take $y_{m,n}^{\mathrm{ID}}$ (2.7) into (2.97) and define $\mathbf{U}_{m,n} = \psi_{m,n}\alpha_{m,n}\left|\mathbf{h}_{m,n}^{H}\mathbf{w}_m\right|^2 + \sum\limits_{j=1}^{M}\sum\limits_{i\in\Omega_j^I}\kappa_{j,i}^{m,n}\psi_{m,n}\alpha_{j,i}$ $\left|\mathbf{h}_{m,n}^{H}\mathbf{w}_j\right|^2 + \psi_{m,n}\sigma_0^2 + \phi_{m,n}\sigma_1^2$, the expression of $\mathrm{MSE}_{m,n}$ is give by

$$
\begin{aligned}
\mathrm{MSE}_{m,n} &= v_{m,n}\mathbf{U}_{m,n}v_{m,n}^{H} - \sqrt{\psi_{m,n}\alpha_{m,n}}\mathbf{w}_m^{H}\mathbf{h}_{m,n}v_{m,n}^{H} - \sqrt{\psi_{m,n}\alpha_{m,n}}v_{m,n}\mathbf{h}_{m,n}^{H}\mathbf{w}_m + 1 \\
&= \left(v_{m,n} - \sqrt{\psi_{m,n}\alpha_{m,n}}\mathbf{w}_m^{H}\mathbf{h}_{m,n}\mathbf{U}_{m,n}^{-1}\right)\mathbf{U}_{m,n}\left(v_{m,n} - \sqrt{\psi_{m,n}\alpha_{m,n}}\mathbf{w}_m^{H}\mathbf{h}_{m,n}\mathbf{U}_{m,n}^{-1}\right)^{-1} \\
&\quad + 1 - \psi_{m,n}\alpha_{m,n}\mathbf{w}_m^{H}\mathbf{h}_{m,n}\mathbf{U}_{m,n}^{-1}\mathbf{h}_{m,n}^{H}\mathbf{w}_m
\end{aligned}
\tag{2.98}
$$

It is easy to know from (2.98) that the $\mathrm{MSE}_{m,n}$ is minimized only when $v_{m,n} = \sqrt{\psi_{m,n}\alpha_{m,n}}\mathbf{w}_m^{H}\mathbf{h}_{m,n}\mathbf{U}_{m,n}^{-1}$. As such, the MMSE can be described as

$$
\begin{aligned}
e_{m,n} &= 1 - \psi_{m,n}\alpha_{m,n}\mathbf{w}_m^{H}\mathbf{h}_{m,n}\mathbf{U}_{m,n}^{-1}\mathbf{h}_{m,n}^{H}\mathbf{w}_m \\
&= \frac{\mathbf{U}_{m,n} - \psi_{m,n}\alpha_{m,n}\left|\mathbf{h}_{m,n}\mathbf{w}_j\right|^2}{\mathbf{U}_{m,n}} \\
&= \frac{1}{1 + \Gamma_{m,n}}
\end{aligned}
\tag{2.99}
$$

and $v_{m,n}^{\mathrm{MMSE}} = \sqrt{\psi_{m,n}\alpha_{m,n}}\mathbf{w}_m^{H}\mathbf{h}_{m,n}\mathbf{U}_{m,n}^{-1}$ is the MMSE receiver.

B: Proof of Proposition 2.1

Prior to starting the proof, we provide some useful lemmas.

Lemma 6.1 (Sylvester's rank inequality, [49], COROLLAR 6.1): For matrix $\mathbf{A}_{m\times n}$ and matrix $\mathbf{B}_{n\times s}$, if $\mathbf{AB} = \mathbf{0}$, then $\mathrm{Rank}(\mathbf{A}) + \mathrm{Rank}(\mathbf{B}) \leq n$.

Lemma 6.2: If $\mathbf{A}$ and $\mathbf{B}$ are two matrices of same size, then $\mathrm{Rank}(\mathbf{A} - \mathbf{B}) \geq \mathrm{Rank}(\mathbf{A}) - \mathrm{Rank}(\mathbf{B})$.

Proof It is known that $\mathrm{Rank}(\mathbf{A}) + \mathrm{Rank}(\mathbf{B}) \geq \mathrm{Rank}(\mathbf{A} + \mathbf{B})$ according to the following equations

$$
\mathrm{Rank}(\mathbf{A}) + \mathrm{Rank}(\mathbf{B}) \geq \mathrm{Rank}\begin{bmatrix}\mathbf{A}\\\mathbf{B}\end{bmatrix} = \mathrm{Rank}\begin{bmatrix}\mathbf{A}+\mathbf{B}\\\mathbf{B}\end{bmatrix}
\tag{2.100}
$$

$$
\geq \mathrm{Rank}\begin{bmatrix}\mathbf{A}+\mathbf{B}\\\mathbf{0}\end{bmatrix} = \mathrm{Rank}(\mathbf{A} + \mathbf{B}).
$$

Thus, we have

$$
\mathrm{Rank}(\mathbf{A} - \mathbf{B}) + \mathrm{Rank}(\mathbf{B}) \geq \mathrm{Rank}(\mathbf{A}).
\tag{2.101}
$$

Obviously, (2.101) is equal to $\text{Rank}(\mathbf{A} - \mathbf{B}) \geq \text{Rank}(\mathbf{A}) - \text{Rank}(\mathbf{B})$.

Lemma 6.3 (Sylvester's rank inequality, [49], COROLLAR 6.1): For matrix $\mathbf{A}_{m \times n}$ and matrix $\mathbf{B}_{n \times s}$, we have $\text{Rank}(\mathbf{AB}) \geq \text{Rank}(\mathbf{A}) + \text{Rank}(\mathbf{B}) - n$.

The Lagrangian function of OP5 can be written as

$$
\begin{aligned}
\mathfrak{L}(\mathbf{W}) = &\sum_{m=1}^{M} \text{tr}(\mathbf{W}_m) - \sum_{m=1}^{M} \sum_{n \in \Omega_m^I} \frac{\lambda_{m,n} \alpha_{m,n}}{\gamma_{m,n}} \text{tr}(\mathbf{H}_{m,n} \mathbf{W}_m) - \sum_{m=1}^{M} \sum_{n \in \Omega_m^I} \sum_{i \in \Omega_m^I} \lambda_{m,n} \kappa_{m,i}^{m,n} \alpha_{m,i} \text{tr}(\mathbf{H}_{m,n} \mathbf{W}_m) \\
&+ \sum_{m=1}^{M} \sum_{n \in \Omega_m^I} \lambda_{m,n}(\sigma_0^2 + \frac{\sigma_1^2}{\rho_{m,n}}) + \sum_{m=1}^{M} \sum_{n \in \Omega_m^I} \sum_{j=1, j \neq m}^{M} tr(\lambda_{m,n} \mathbf{H}_{m,n} \mathbf{W}_j) \\
&+ \sum_{m=1}^{M} \sum_{n \in \Omega_m^E} \tau_{m,n} \left[\frac{Q_{m,n}}{1 - \psi_{m,n}} - \sum_{j=1}^{M} \text{tr}(\mathbf{H}_{m,n} \mathbf{W}_j) \right] - \sum_{m=1}^{M} \mathbf{Z}_m \mathbf{W}_m,
\end{aligned}
\tag{2.102}
$$

where $\lambda_{m,n}, \tau_{m,n}, \mathbf{Z}_m$ are the Lagrange multipliers of the C7, C8, and C10, respectively. Since OP5 is convex, Slater's condition and Karush-Kuhn-Tucher (KKT) conditions are the sufficient and necessary optimality conditions. Then, by making use of KKT conditions, we have

$$
\sum_{j=1}^{M} \text{tr}\left(\mathbf{H}_{m,n} \mathbf{W}_j^*\right) - \frac{Q_{m,n}}{1 - \psi_{m,n}} = 0, n \in \Omega_m^E \tag{2.103a}
$$

$$
\mathbf{Z}_m^* \mathbf{W}_m^* = \mathbf{0}, \forall m, \tag{2.103b}
$$

$$
\nabla_{\mathbf{W}_m^*} \mathfrak{L} = \mathbf{I} - \lambda_{m,n}^* \mathbf{H}_{m,n}^H \left(\frac{\alpha_{m,n}}{\gamma_{m,n}} - \sum_{i \in \Omega_m^I} \kappa_{m,i}^{m,n} \alpha_{m,i} \right)
$$
$$
- \sum_{m=1}^{M} \sum_{n \in \Omega_m^E} \tau_{m,n}^* \mathbf{H}_{m,n}^H - \mathbf{Z}_m^* = \mathbf{0} \tag{2.103c}
$$

$$
\mathbf{W}_m^* \succeq \mathbf{0}, \ \lambda_{m,n}^* \geq 0, \ \tau_{m,n}^* \geq 0, \ \mathbf{Z}_m^* \succeq \mathbf{0} \tag{2.103d}
$$

From (2.103a), we can deduce that $\mathbf{W}_m^* \neq \mathbf{0}$ due to $\frac{Q_{m,n}}{1 - \psi_{m,n}} > 0$. That is,

$$
\text{Rank}(\mathbf{W}_m^*) \geq 1, \forall m \tag{2.104}
$$

Then, using *Lemma 6.1*, we have from (2.103b) that

$$
\text{Rank}(\mathbf{Z}_m^*) \leq N_t - 1, \forall m \tag{2.105}
$$

Next, according to the *Lemma 6.2*, we further obtain from (2.103c) that

$$\text{Rank}(\mathbf{Z}_m^*) = \text{Rank}(\mathbf{I} - \varpi_{m,n}\mathbf{H}_{m,n}^H)$$
$$\geq N_t - 1 \tag{2.106}$$

where $\varpi_{m,n} = \frac{\lambda_{m,n}^*\alpha_{m,n}}{\gamma_{m,n}} - \sum_{i\in\Omega_m^I}\lambda_{m,n}^*\kappa_{m,i}^{m,n}\alpha_{m,i} - \sum_{m=1}^{M}\sum_{n\in\Omega_m^E}\tau_{m,n}^*$. Combing (2.105) and (2.106), it is obvious that $\text{Rank}(\mathbf{Z}_m^*) = N_t - 1, \forall m$. In term of (2.103b), it is obtained that $\text{Rank}(\mathbf{Z}_m^*\mathbf{W}_m^*) = 0$. Moreover, based on *Lemma 6.3*, it is clear that $\text{Rank}(\mathbf{Z}_m^*\mathbf{W}_m^*) \geq \text{Rank}(\mathbf{Z}_m^*) + \text{Rank}(\mathbf{W}_m^*) - N_t$. Thus, we have

$$\text{Rank}(\mathbf{W}_m^*) \leq \text{Rank}(\mathbf{Z}_m^*\mathbf{W}_m^*) - \text{Rank}(\mathbf{Z}_m^*) + N_t$$
$$= 0 - (N_t - 1) + N_t = 1 \tag{2.107}$$

According to (2.104) and (2.107), we can easily confirm that $\text{Rank}(\mathbf{W}_m^*) = 1$. Therefore, we get the Proposition 2.1, which proves that SDR processing from OP4 to OP5 is tight. That is, when the optimal solution of OP5 is found, the corresponding optimal solution of OP4 also can be obtained. Thus, we get the Proposition 2.1.

C: Proof of Theorem 2.1

For convenience of expression, let $\{\mathbf{W}_m^{(t)}, x_{m,n}^{(t)}, y_{m,n}^{(t)}, t = 1, 2, \cdots\}$ be the solutions of problem (2.54) in the tth iteration. Before performing iteration, we first set a feasible initial value $\mathbf{w}_m^{(0)}$. Based on the initial value $\mathbf{W}_m^{(0)} = \mathbf{w}_m^{(0)}(\mathbf{w}_m^{(0)})^H$, we can get $\tilde{y}_{m,n}^{(1)}$. Then, the solutions of the first round of iteration $\{\mathbf{W}_m^{(1)}, x_{m,n}^{(1)}, y_{m,n}^{(1)}\}$ can be obtained accordingly. Finally, we need to ensure the feasibility of the initial value base on the constraint that $x_{m,n}$ and $y_{m,n}$ are positive. Due to the form of the objective function (2.54a), the optimal solution $y_{m,n}^{(t)}$ is required to satisfy

$$y_{m,n}^{(t)} \leq \tilde{y}_{m,n}^{(t)}. \tag{2.108}$$

Then, according to the constraint (2.53), we have

$$e^{\tilde{y}_{m,n}^{(t+1)}} = \sum_{j=1}^{M}\sum_{i=1}^{N_j}\varphi_{j,i}^{m,n}\alpha_{j,i}\left(\hat{\mathbf{h}}_{m,n}^H\mathbf{W}_j^{(t)}\hat{\mathbf{h}}_{m,n} + 2\varepsilon_{m,n}\left\|\mathbf{W}_j^{(t)}\hat{\mathbf{h}}_{m,n}\right\|\right) + \sigma_0^2 + \frac{\sigma_1^2}{\rho_{m,n}}$$
$$= e^{\tilde{y}_{m,n}^{(t)}}\left(y_{m,n}^{(t)} - \tilde{y}_{m,n}^{(t)} + 1\right)$$
$$\leq e^{y_{m,n}^{(t)}}. \tag{2.109}$$

Combine (2.108) and (2.109), it is obtained that

$$\tilde{y}_{m,n}^{(t+1)} \leq y_{m,n}^{(t)} \leq \tilde{y}_{m,n}^{(t)}, \tag{2.110}$$

i.e., $\{\tilde{y}_{m,n}^{(t)}\}$ is monotonically decreasing. Moreover, as mentioned in (2.97), we have $e^{\tilde{y}_{m,n}^{(t)}} \geq e^{y_{m,n}^{(t)}} \geq \sigma_0^2 + \frac{\sigma_1^2}{\rho_{m,n}}$, which means $\{\tilde{y}_{m,n}^{(t)}\}$ has a lower bound. According to the monotone convergence theorem, sequence $\{\tilde{y}_{m,n}^{(t)}\}$ would converge. Once $\{\tilde{y}_{m,n}^{(t)}\}$ converges, the solution $\{y_{m,n}^{(t)}\}$ would also converge due to (2.108). Because of $y_{m,n}^{(t+1)} \leq y_{m,n}^{(t)}$, the solutions $\{x_{m,n}^{(t)}, y_{m,n}^{(t)}\}$ obtained in the tth iteration are feasible in the $(t+1)$th iteration. Then, due to the form of the objective function (2.54a), we have

$$x_{m,n}^{(t+1)} \geq x_{m,n}^{(t)}, \tag{2.111}$$

which means $\{x_{m,n}^{(t)}\}$ is monotonically increasing. Since $\mathbf{W}_m$ is limited by $P_{\max}$, $x_{m,n}^{(t)}$ is bounded, i.e., $\{x_{m,n}^{(t)}\}$ has an upper bound. Based on the monotone convergence theorem, sequence $\{x_{m,n}^{(t)}\}$ would converge. Thus, the solutions $\{x_{m,n}^{(t)}, y_{m,n}^{(t)}, t = 1, 2, \cdots\}$ generated by the proposed Algorithm 2.5 would converge. As a result, the optimal solution $\{\mathbf{W}_m^{(t)}\}$ to problem (2.54) would converge in the iterations.

D: Proof of Theorem 2.2

For the post-SIC received signal for ID $y_{m,n}^{\text{ID}}$, the MSE associated with the $\text{UE}_{m,n}$ is given by

$$\text{MSE}_{m,n} = \mathbb{E}\left\{ \left(v_{m,n} y_{m,n}^{\text{ID}} - s_{m,n} \right) \left(v_{m,n} y_{m,n}^{\text{ID}} - s_{m,n} \right)^H \right\}, \tag{2.112}$$

where $v_{m,n}$ denotes the ID receiver at the $\text{UE}_{m,n}$. Substituting (2.81) into (2.112), $\text{MSE}_{m,n}$ can be expressed as

$$\text{MSE}_{m,n} = v_{m,n} \left(\sum_{j=1}^{M} \sum_{i=1}^{N_j} \rho_{m,n} \kappa_{j,i}^{m,n} \alpha_{j,i} \delta_{m,n} \mathbf{h}_{m,n}^H \mathbf{w}_j \mathbf{w}_j^H \mathbf{h}_{m,n} + \rho_{m,n} \sigma_0^2 + \sigma_1^2 \right) v_{m,n}^H$$
$$- 2 \text{Re}\left\{ \sqrt{\rho_{m,n} \alpha_{m,n} \delta_{m,n}} \mathbf{w}_j^H \mathbf{h}_{m,n} v_{m,n}^H \right\} + 1, \tag{2.113}$$

where

$$\kappa_{j,i}^{m,n} = \begin{cases} \eta_{m,n}, & \text{if } j = m \text{ and } i > n \\ 1, & \text{otherwise.} \end{cases} \tag{2.114}$$

Due to partial CSI at the BS, i.e., $\mathbf{h}_{m,n} = \sqrt{\chi_{m,n}} \hat{\mathbf{h}}_m + \mathbf{e}_{m,n}$, it is difficult to obtain an exact expression for the average rate $\bar{R}_{m,n}$ at the $\text{UE}_{m,n}$. To this end, we consider a lower bound on average rate according to the relationship between MSE and average rate [46]

$$\mathbb{E}\left\{ -\log_2\left(\text{MSE}_{m,n} \right) \right\} \geq -\log_2\left(\mathbb{E}\left\{ \text{MSE}_{m,n} \right\} \right) = \bar{R}_{m,n}. \tag{2.115}$$

Thus, we transform the original objective function into maximizing the lower bound on the average WSR, which is equivalent to minimizing the average MSE. Without loss of generality, we consider the average MSE of the $UE_{m,n}$, which is given by

$$\overline{MSE}_{m,n} = 1 + v_{m,n}\left(\rho_{m,n}\sigma_0^2 + \sigma_1^2\right)v_{m,n}^H - 2\,\mathrm{Re}\left\{\sqrt{\rho_{m,n}\alpha_{m,n}\delta_{m,n}\chi_{m,n}}\,\mathbf{w}_m^H\hat{\mathbf{h}}_m v_{m,n}^H\right\}$$
$$+ v_{m,n}\left(\sum_{j=1}^{M}\sum_{i=1}^{N_j}\rho_{m,n}\kappa_{j,i}^{m,n}\alpha_{j,i}\delta_{m,n}\left(\chi_{m,n}\hat{\mathbf{h}}_m^H\mathbf{w}_j\mathbf{w}_j^H\hat{\mathbf{h}}_m + \Psi_j^{m,n}\right)\right)v_{m,n}^H,$$
(2.116)

where

$$\Psi_j^{m,n} = \mathbb{E}\left\{\mathbf{e}_{m,n}^H\mathbf{w}_j\mathbf{w}_j^H\mathbf{e}_{m,n}\right\} = \mathrm{tr}\left(\mathbf{w}_j\mathbf{w}_j^H\left(1 - \chi_{m,n}\right)\right) = \left(1 - \chi_{m,n}\right)\|\mathbf{w}_j\|_F^2.$$
(2.117)

Then, letting $O_{m,n} = \sum_{j=1}^{M}\sum_{i=1}^{N_j}\rho_{m,n}\kappa_{j,i}^{m,n}\alpha_{j,i}\delta_{m,n}\left(\chi_{m,n}\hat{\mathbf{h}}_m^H\mathbf{w}_j\mathbf{w}_j^H\hat{\mathbf{h}}_m + \Psi_j^{m,n}\right) + \rho_{m,n}\sigma_0^2 + \sigma_1^2$, the average MSE of the $UE_{m,n}$ can be rewritten as

$$\overline{MSE}_{m,n} = v_{m,n}O_{m,n}v_{m,n}^H - 2\,\mathrm{Re}\left\{\sqrt{\rho_{m,n}\alpha_{m,n}\delta_{m,n}\chi_{m,n}}\,\mathbf{w}_m^H\hat{\mathbf{h}}_m v_{m,n}^H\right\} + 1$$
$$= \left(v_{m,n} - \sqrt{\rho_{m,n}\alpha_{m,n}\delta_{m,n}\chi_{m,n}}\,\mathbf{w}_m^H\hat{\mathbf{h}}_m O_{m,n}^{-1}\right)O_{m,n}\left(v_{m,n} - \sqrt{\rho_{m,n}\alpha_{m,n}\delta_{m,n}\chi_{m,n}}\,\mathbf{w}_m^H\hat{\mathbf{h}}_m O_{m,n}^{-1}\right)^H$$
$$+ 1 - \rho_{m,n}\alpha_{m,n}\delta_{m,n}\chi_{m,n}\mathbf{w}_m^H\hat{\mathbf{h}}_m O_{m,n}^{-1}\hat{\mathbf{h}}_m^H\mathbf{w}_m.$$
(2.118)

Checking (2.118), it is found that the average MSE is minimized only when the receiver $v_{m,n} = \sqrt{\rho_{m,n}\alpha_{m,n}\delta_{m,n}\chi_{m,n}}\,\mathbf{w}_m^H\hat{\mathbf{h}}_m O_{m,n}^{-1}$, namely the MAMSE receiver. The proof is completed.

References

1. Chen X, Ng DWK, Yu W, Larsson EG, Al-Dhahir N, Schober R (2020) Massive access for 5G and beyond. IEEE J Sel Area Commun (99):1–24
2. Yu G, Chen X, Ng DWK (2019) Low-cost design of massive access for cellular internet of things. IEEE Trans Commun 67(11):8008–8020
3. Ding Z, Lei X, Karagiannidis GK, Schober R, Yuan J, Bhargava VK (2017) A survey on non-orthogonal multiple access for 5G networks: research challenges and future trends. IEEE J Sel Areas Commun 35(10):2181–2195
4. Chen X, Zhang Z, Zhong C, Ng DWK (2017) Exploiting multiple-antenna for non-orthogonal multiple access. IEEE J Sel Areas Commun 35(10):2207–2220
5. Sun X, Yang N, Yan S, Ding Z, Ng DWK, Shen C, Zhong Z (2018) Joint beamforming and power allocation in downlink NOMA multiuser MIMO networks. IEEE Trans Wirel Commun 17(8):5367–5381

6. Chen X, Jia R (2018) Exploiting rateless coding for massive access. IEEE Trans Vehic Technol 67(11):11253–11257
7. Chen X, Jia R, Ng DWK (2019) On the design of massive non-orthogonal multiple access with imperfect successive interference cancellation. IEEE Trans Commun 67(3):2539–2551
8. Na W, Park J, Lee C, Park K, Kim J, Cho S (2018) Energy-efficient mobile charging for wireless power transfer in internet of things networks. IEEE Internet of Things J 5(1):79–92
9. Zhou X, Zhang R, Ho CK (2013) Wireless information and power transfer: architecture design and rate-energy tradeoff. IEEE Trans Commun 61(11):4754–4767
10. Qi Q, Chen X (2019) Wireless powered massive access for cellular internet of things with imperfect SIC and non-linear EH. IEEE Internet of Things J 6(2):3110–3120
11. Chen X, Zhang Z, Chen H-H, Zhang H (2015) Enhancing wireless information and power transfer by exploiting multi-antenna techniques. IEEE Commun Mag 53(4):133–141
12. Tian F, Chen X (2019) Multiple-antenna techniques in nonorthogonal multiple access: a review. Front Inform Technol Electron Eng 20(12):1665–1697
13. Chen X, Yuen C, Zhang Z (2014) Wireless energy and information transfer tradeoff for limited feedback multi-antenna systems with energy beamforming. IEEE Trans Vehic Technol 63(1):407–412
14. Chen X, Wang X, Chen X (2013) Energy-efficient optimization for wireless information and power transfer in large-scale MIMO systems employing energy beamforming. IEEE Wirel Commun Lett 2(6):667–670
15. Gesbert D, Hanly S, Huang H, Shitz SS, Simeone O, Yu W (2010) Multi-cell MIMO cooperative networks: a new look at interference. IEEE J Sel Areas Commun 28(9):1380–1408
16. Chen X, Zhang Z, Chen H-H (2010) On distributed antenna system with limited feedback precoding-opportunities and challenges. IEEE Wirel Commun 17(2):80–88
17. Chen X, Zhang Z, Zhong C, Jia R, Ng DWK (2018) Fully non-orthogonal communication for massive access. IEEE Trans Commun 66(4):1717–1731
18. Qi Q, Chen X, Ng DWK (2020) Robust beamforming for NOMA-based cellular massive IoT with SWIPT. IEEE Trans Sig Process (99):1–15 (2020)
19. Wang J, Palomar DP (2009) Worst-case robust MIMO transmission with imperfect channel knowledge. IEEE Trans Sig Process 57(8):3086–3100
20. Liao J, Khandaker MRA, Wong K-K (2016) Robust power-splitting SWIPT beamforming for broadcast channels. IEEE Sig Process Lett 20(1):181–184
21. Zhou F, Chu Z, Sun H, Hu RQ, Hanzo L (2018) Artificial noise aided secure cognitive beamforming for cooperative MISO-NOMA using SWIPT. IEEE J Sel Areas Commun 36(4):918–931
22. Sun H, Zhou F, Hu RQ, Hanzo L (2019) Robust beamforming design in a NOMA cognitive radio network relying on SWIPT. IEEE J Sel Areas Commun 37(1):142–155
23. Wang F, Peng T, Huang Y, Wang X (2015) Robust transceiver optimization for power-splitting based downlink MISO SWIPT systems. IEEE Sig Process Lett 22(9):1492–1496
24. Zhou F, Li Z, Cheng J, Li Q, Si J (2017) Robust AN-aided beamforming and power splitting design for secure MISO cognitive radio with SWIPT. IEEE Trans Wirel Commun 16(4):2450–2464
25. Roger S, Calabuig D, Cabrejas J, Monserrat JF (2014) Multi-user noncoherent detection for downlink MIMO communication. IEEE Sig Process Lett 21(10):1225–1229
26. Raghavan V, Kotecha JH, Sayeed AM (2010) Why does the kronecker model result in misleading capacity estimates? IEEE Trans. Inf. Theory 56(10):4843–4864
27. Sun H, Xie B, Hu RQ, Wu G (2016) Non-orthogonal multiple access with SIC error propagation in downlink wireless MIMO networks. In: Proceeding of IEEE VTC-Fall, invited paper, pp 1–5
28. Zhu G, Zhong C, Suraweera H, Karagiannids G, Zhang Z, Tsiftsis T (2015) Wireless information and power transfer in relay systems with multiple antennas and interference. IEEE Trans Commun 63(4):1400–1418
29. Chen X, Ng DWK, Chen H-H (2016) Secrecy wireless information and power transfer: challenges and opportunities. IEEE Wirel Commun 23(2):54–61

30. Boshkovska E, Ng DWK, Zlatanov N, Schober R (2015) Practical non-linear energy harvesting model and resource allocation for SWIPT systems. IEEE Commun Lett 19(12):2082–2085
31. Valenta CR, Durgin GD (2014) Harvesting wireless power: survey of energy-harvester conversion efficiency in far-field, wireless power transfer systems. IEEE Microw Mag 15(4):108–120
32. Le T, Mayaram K, Fiez T (2008) Efficient far-field radio frequency energy harvesting for passively powered sensor networks. IEEE J Solid-State Circ 15(4):1287–1302
33. Shi Q, Razaviyayn M, Luo Z-Q, He C (2011) An iteratively weighted MMSE approach to distributed sum-utility maximization for a MIMO interfering broadcast channel. IEEE Trans Sig Process 59(9):431–4340
34. Ye Y (1997) Interior point algorithms: theory and analysis. Wiley, New York
35. Boyd S, Vandenberghe L (2004) Convex Optimization. Cambridge University Press, Cambridge, UK
36. Grant M, Boyd S. CVX: Matlab software for disciplined convex programming. http://cvxr.com/cvx
37. Hardy GH, Littlewood JE, Polya G (1952) Inequalities. Cambridge University Press, Cambridge, UK
38. Pataki G (1998) On the rank of extreme matrices in semidefinite programs and the multiplicity of optimal eigenvalues. Math Oper Res 23:339–358
39. Ben-Tal A, Nemirovski A (2001) Lectures on modern convex optimization: analysis, algorithms, and engineering applications. SIAM, MPS-SIAM Series on Optimization, Philadelphia, PA, USA
40. Krishnamurthy S, Tzoneva R (2013) Investigation on the impact of the penalty factors over solution of the dispatch optimization problem. In: Proceeding of IEEE ICIT, Cape Town, pp 851–860
41. Chen X (2019) Massive access for cellular internet of things theory and technique. Springer, Germany
42. Shao X, Chen X, Zhong C, Zhao J, Zhang Z (2019) A unified sesign of massive access for cellular internet of things. IEEE Internet of Things J 6(2):3934–3947
43. Yang Z, Ding Z, Fan P, Karagiannidis GK (2016) On the performance of non-orthogonal multiple access systems with partial channel information. IEEE Trans Commun 64(2):654–667
44. Islam SMR, Avazov N, Dobre OA, Kwak K-S (2017) Power-domain non-orthogonal multiple access (NOMA) in 5G: potentials and challenges. IEEE Commun Surv Tutor 19(2):721–742
45. Wang KY, So AMC, Chang TH, Ma WK, Chi CY (2014) Outage constrained robust transmit optimization for multiuser MISO downlinks: tractable approximations by conic optimization. IEEE Trans Sig Process 62(21):5690–5705
46. Fritzsche R, Fettweis GP (2013) Robust sum rate maximization in the multi-cell mu-mimo downlink. In: Proceeding of IEEE WCNC, pp 3180–3184
47. Jia R, Chen X, Zhong C, Ng DWK, Lin H, Zhang Z (2019) Design of non-orthogonal beamspace multiple access for cellular internet-of-things. IEEE J Sel Top Sig Process 13(3):538–552
48. 3GPP, Coordinated multi-point operation for LTE physical layer aspects (Release 11) (2011)
49. Matsaglia G, Styan GPH (1974) Equalities and inequalities for ranks of matrices. Lin Multilin Algebra 2:269–292

Chapter 3
Convergence of Energy and Computation in B5G Cellular Internet of Things

Abstract In B5G cellular IoT, energy supply and data aggregation of a massive number of devices are two vitally challenging issues. To address these challenges, we propose a wireless powered MIMO AirComp design framework. Firstly, WPT is utilized to charge massive IoT devices simultaneously by exploiting the open nature of wireless broadcast channels. Then, AirComp is adopted to reduce latency of massive data aggregation via exploring the superposition property of wireless multiple-access channels. To realize efficient convergence of energy and computation in B5G cellular IoT under practical but adverse conditions, a robust design algorithm is provided by jointly optimizing beamforming of both WPT and AirComp. Finally, extensive simulation results validate the robustness and effectiveness of the proposed algorithm over the baseline ones.

3.1 Introduction

Future B5G cellular IoT is converting from a data-centric network to a computation-centric one [1]. The advanced information processing technologies, such as artificial intelligence and data mining, will provide ubiquitous computing and intelligent services to realize analysis and processing of massive data from IoT devices, which means that we may be more concerned about the computation results of the data rather than the individual data itself in the future [2]. Thus, the conventional approach of *transmit-then-compute* is quite unsuitable for B5G cellular IoT with massive access due to the excessively high latency and low spectrum utilization [3]. To address this issue, a promising solution called *over-the-air computation* (AirComp) was proposed in [4], which exploited the superposition property of wireless multiple-access channels to compute a class of *nomographic functions* of distributed data from IoT devices via concurrent transmission [5, 6].

In contrast of rate-centric wireless communications where simultaneous transmissions cause interference, the accuracy of computation enabled by AirComp may be improved as the number of simultaneous IoT devices increases [7]. Specifically, AirComp that utilizes interference from concurrent transmission for functional

© The Author(s), under exclusive license to Springer Nature Singapore Pte Ltd. 2020
X. Chen and Q. Qi, *Convergence of Energy, Communication and Computation in B5G Cellular Internet of Things*, SpringerBriefs in Electrical and Computer Engineering, https://doi.org/10.1007/978-981-15-4140-7_3

computation can significantly decrease the data aggregation latency by a factor equal to the number of IoT devices. In the coming B5G era, AirComp combining with multiple-input and multiple-output (MIMO) techniques, namely MIMO AirComp, exploits spatial degrees of freedom provided by large-scale antenna arrays to spatially multiplex multi-function computation, and can further reduce computation errors by using spatial beamforming [8]. Hence, the key of the MIMO AirComp design lies in beamforming optimization in terms of minimizing the computation distortion of targeted functions. In [9], a beamforming optimization scheme for MIMO AirComp in sensors network was well analyzed and designed. The authors in [10] designed a reduced-dimension MIMO AirComp framework consisting of beamforming optimization for clustered IoT networks. Note that most of existing works about beamforming design for MIMO AirComp systems are all based on the assumption of perfect channel state information (CSI), which is impractical in B5G cellular IoT with massive access. Thus, it is necessary to design a practical MIMO AirComp system with partial CSI.

The other critical issue of B5G cellular IoT is energy supply for massive IoT devices. Due to the high cost of battery replacement as well as environmental factors, the conventional battery replacement-based solutions may no longer be feasible for B5G cellular IoT. To solve it, a technique called wireless power transfer (WPT) based on radio-frequency (RF) signals was proposed and energy beamforming was accordingly introduced to enhance the efficiency of WPT over fading channels [11, 12]. Especially, WPT can charge a massive number IoT devices simultaneously due to the open nature of wireless broadcast channels [13]. Similarly, the key of WPT in B5G cellular IoT is the design of beamforming based on partial CSI [14–16].

To address the above challenges in B5G cellular IoT, this chapter advocates the robust convergence of energy and computation, namely wireless powered MIMO AirComp. The contributions of this chapter are two-fold:

1. We propose a comprehensive design framework for the computation-centric B5G cellular IoT, including WPT for battery charging in the downlink and MIMO-AirComp for data aggregation in the uplink.
2. We provide a robust algorithm to improve the overall performance of energy and computation convergence by jointly optimizing beamforming of WPT and AirComp.

The rest of this chapter is organized as follows. Section 3.2 offers a concise introduction of B5G cellular IoT with wireless powered AirComp. Section 3.3 focuses on the design of a robust beamforming algorithm. Section 3.4 presents several simulation results to validate the effectiveness and robustness of the proposed algorithm. Finally, Sect. 3.5 summarizes the chapter.

Notations: We use bold upper (lower) letters to denote matrices (column vectors), $(\cdot)^H$ to denote conjugate transpose, $\| \cdot \|$ and $\| \cdot \|_F$ to respectively denote the L_2-norm of a vector and the F-norm of a matrix, $| \cdot |$ to denote the absolute value, Re$\{\cdot\}$ to denote the real parts of a complex number, E$\{\cdot\}$ to denote the expectation value, tr$(\cdot)$ to denote trace of a matrix, and Rank$(\cdot)$ to denote rank of a matrix, $\mathbb{R}$ and $\mathbb{C}$ to denote the real domain and complex domain, respectively.

3.2 System Model

Let us consider a computation-centric B5G cellular IoT network operated in time division duplex mode, where a BS equipped with N_b antennas plays two roles, i.e., a power beacon in the downlink and a data fusion center in the uplink for K multi-modal IoT user equipments (UEs) equipped with N_u antennas each, c.f. Fig. 3.1. At the first half duration of a time slot of length T, the BS utilizes the WPT technique to charge IoT UEs via energy beamforming. Without loss of generality, the harvested energy at the kth UE can be expressed as

$$Q_k = \theta_k \left\| \mathbf{H}_k^H \mathbf{v} \right\|^2 T/2, \tag{3.1}$$

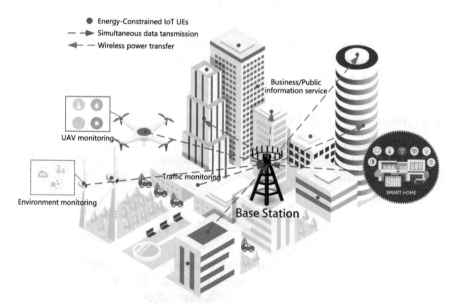

(a) A model of computation-centric B5G cellular IoT network

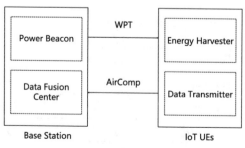

(b) System structure

Fig. 3.1 A framework of computation-centric B5G cellular IoT network

where $\mathbf{H}_k \in \mathbb{C}^{N_b \times N_u}$ denotes the MIMO channel matrix between the BS and the kth UE. It is reasonably assumed that $\mathbf{H}_k$ keeps unchanged during a time slot, but independently fades over time slots. θ_k denotes the power conversion efficiency and $\mathbf{v} \in \mathbb{C}^{N_b \times 1}$ denotes an energy beam in the downlink.

In the second half duration of a time slot, all IoT UEs transmit a set of multi-modal data to the BS simultaneously with the harvested energy. Note that each IoT UE logs data of L heterogeneous time-varying parameters from the environment or human, which generates a vector symbol $\mathbf{d}_k = \left[d_{k,1}, d_{k,2}, \ldots, d_{k,L}\right]^T \in \mathbb{R}^{L \times 1}, k = 1, \ldots, K$, in each slot, where $d_{k,l}$ is the measured value of the parameter l at the kth UE. For the data aggregation, the BS engages itself in computing L nomographic functions [5, 6], such that

$$y_l = f_l \left(\sum_{k=1}^{K} g_{k,l} \left(d_{k,l} \right) \right), l = 1, \ldots, L, \tag{3.2}$$

where $f_l(\cdot)$ and $g_{kl}(\cdot)$ represent post-processing functions at the BS and pre-processing functions at the IoT UEs, respectively (see Table 3.1 for examples). Let $\mathbf{s}_k = \left[g_{k,1}\left(d_{k,1}\right), g_{k,2}\left(d_{k,2}\right), \ldots, g_{k,L}\left(d_{k,L}\right)\right]^T$ denotes the transmitted signal after pre-processing at the kth IoT UE, where $\mathrm{E}\left\{\mathbf{s}_k \mathbf{s}_k^H\right\} = \mathbf{I}$. Due to the one-to-one mapping between $\mathbf{s} = \sum_{k=1}^{K} \mathbf{s}_k$ and $\mathbf{y} = [y_1, y_2, \ldots, y_L]^T$ in (3.2), we take an accurate $\mathbf{s}$ at the BS as a target-function signal. To minimize the distortion of the function values caused by channel fading and noise, it is desired to perform beamforming at the IoT UEs and the BS jointly. Thus, the received signal at the BS is given by

$$\hat{\mathbf{s}} = \mathbf{Z}^H \sum_{k=1}^{K} \mathbf{H}_k \mathbf{W}_k \mathbf{s}_k + \mathbf{Z}^H \mathbf{n}, \tag{3.3}$$

where $\mathbf{Z} \in \mathbb{C}^{N_b \times L}$ denotes the receive beamforming matrix at the BS, $\mathbf{W}_k \in \mathbb{C}^{N_u \times L}$ denotes the transmit beamforming matrix at the kth IoT UE, and $\mathbf{n}$ is the additive white Gaussian noise (AWGN) vector with the distribution $\mathscr{CN}\left(\mathbf{0}, \sigma_n^2 \mathbf{I}\right)$. Since the

Table 3.1 Some examples of nomographic functions

Functions	g_k	f	y
Arithmetic mean	$g_k = d_k$	$f = 1/K$	$y = \frac{1}{K} \sum_{k=1}^{K} d_k$
Weighted sum	$g_k = \vartheta_k d_k$	$f = 1$	$y = \sum_{k=1}^{K} \vartheta_k d_k$
Geometric mean	$g_k = \ln(d_k)$	$f = \exp(\cdot)$	$y = \left(\prod_{k=1}^{K} d_k\right)^{1/K}$
Polynomial	$g_k = \vartheta_k d_k^{\beta_k}$	$f = 1$	$y = \sum_{k=1}^{K} \vartheta_k d_k^{\beta_k}$
Euclidean norm	$g_k = d_k^2$	$f = \sqrt{(\cdot)}$	$y = \sqrt{\sum_{k=1}^{K} d_k^2}$

transmit power of each IoT UE stems from the harvested energy in the downlink, we have

$$\text{tr}\left(\mathbf{W}_k \mathbf{W}_k^H\right) \leq \frac{\theta_k T/2}{T/2} \left\|\mathbf{H}_k^H \mathbf{v}\right\|^2, k = 1, ..., K. \tag{3.4}$$

In general, the performance of AirComp is measured by the mean-squared error (MSE) between $\mathbf{s}$ and $\hat{\mathbf{s}}$, which can be expressed as

$$\text{MSE}\left(\hat{\mathbf{s}}, \mathbf{s}\right) = \text{E}\left\{\text{tr}\left(\left(\hat{\mathbf{s}} - \mathbf{s}\right)\left(\hat{\mathbf{s}} - \mathbf{s}\right)^H\right)\right\}. \tag{3.5}$$

Substituting (3.3) into (3.5), the computation distortion can be replaced by the MSE function of receive and transmit beams, which is given by

$$\text{MSE}\left(\mathbf{Z}, \mathbf{W}_k\right) = \sum_{k=1}^{K} \left\|\mathbf{Z}^H \mathbf{H}_k \mathbf{W}_k - \mathbf{I}\right\|_F^2 + \sigma_n^2 \left\|\mathbf{Z}\right\|_F^2. \tag{3.6}$$

As seen from (3.4) to (3.6), the system performance is jointly affected by energy beam $\mathbf{v}$ and receive beam $\mathbf{Z}$ at the BS, and transmit beams $\mathbf{W}_k, k = 1, ..., K$, at the IoT UEs. It is known that beamforming design is closely linked to the CSI. However, in the B5G cellular IoT with massive access, it is only able to obtain partial or even no CSI. In other words, it is necessary to take the uncertainty of CSI into consideration for beamforming design, namely robust beamforming. To characterize the CSI uncertainty, we adopt a deterministic imperfect CSI model [17]. In particular, the real channel matrix $\mathbf{H}_k$ related to the kth IoT UE can be expressed as

$$\mathscr{H}_k \triangleq \left\{\mathbf{H}_k = \hat{\mathbf{H}}_k + \boldsymbol{\Delta}_k \,\middle|\, \|\boldsymbol{\Delta}_k\|_F \leq \varepsilon_k\right\}, \tag{3.7}$$

where $\hat{\mathbf{H}}_k$ is the obtained channel matrix and $\boldsymbol{\Delta}_k$ is the channel error matrix, whose Frobenius norm is bounded by a given radius ε_k. To improve the performance of computation-centric B5G cellular IoT, it is desired to jointly design beamforming of WPT and AirComp in the presence of partial CSI, namely robust convergence of energy supply and data aggregation.

3.3 Problem Formulation and Optimization Solution

In this section, we design a worst-case robust beamforming algorithm for computation-centric B5G cellular IoT with the goal of minimizing the computation errors under the transmission power constraints, which can be formulated as the following optimization problem:

$$\min_{\mathbf{v},\mathbf{Z},\{\mathbf{W}_k\}} \max_{\mathbf{H}_k \in \mathcal{H}_k} \sum_{k=1}^{K} \left\| \mathbf{Z}^H \mathbf{H}_k \mathbf{W}_k - \mathbf{I} \right\|_F^2 + \sigma_n^2 \left\| \mathbf{Z} \right\|_F^2 \tag{3.8a}$$

$$\text{s.t. } \left\| \mathbf{W}_k \right\|_F^2 \le \min_{\mathbf{H}_k \in H_k} \theta_k \left\| \mathbf{H}_k^H \mathbf{v} \right\|^2, \tag{3.8b}$$

$$\left\| \mathbf{v} \right\|^2 \le P_{\max}, \tag{3.8c}$$

where $P_{\max}$ is the maximum transmit power budget at the BS. It is known that, by introducing an auxiliary variable η_k, $\forall k$, the min–max problem (3.8) is equivalent to

$$\min_{\mathbf{Z},\{\mathbf{W}_k\},\{\eta_k\}} \sum_{k=1}^{K} \eta_k + \sigma_n^2 \left\| \mathbf{Z} \right\|_F^2 \tag{3.9a}$$

$$\text{s.t. } (5.11b),$$

$$\left\| \mathbf{W}_k \right\|_F^2 \le \theta_k \left\| \left(\hat{\mathbf{H}}_k + \boldsymbol{\Delta}_k \right)^H \mathbf{v} \right\|^2, \tag{3.9b}$$

$$\left\| \mathbf{Z}^H \left(\hat{\mathbf{H}}_k + \boldsymbol{\Delta}_k \right) \mathbf{W}_k - \mathbf{I} \right\|_F^2 \le \eta_k, \tag{3.9c}$$

$$\eta_k \ge 0, \tag{3.9d}$$

$$\forall \boldsymbol{\Delta}_k : \left\| \boldsymbol{\Delta}_k \right\|_F \le \varepsilon_k.$$

However, it is still non-convex due to constraint (3.9b) and (3.9c) both involving channel uncertainty. First, we introduce an useful lemma to handle the constraint (3.9b).

Lemma 3.1 *Define a function $f(\mathbf{X}) = \text{Re}\{tr(\mathbf{X}\mathbf{Y})\}$, where $\mathbf{X}$ is a norm-bounded matrix variable with a given radius δ, i.e., $\|\mathbf{X}\|_F \le \delta$ and matrix $\mathbf{Y}$ is known. The range of $f(\mathbf{X})$ is given by*

$$-\delta \|\mathbf{Y}\|_F \le f(\mathbf{X}) \le \delta \|\mathbf{Y}\|_F. \tag{3.10}$$

Proof Please refer to Appendix A.

Then, we deal with the term $\left\| \left(\hat{\mathbf{H}}_k + \boldsymbol{\Delta}_k \right)^H \mathbf{v} \right\|^2$ in the constraint (3.9b). Specifically, we have

$$\left\| \left(\hat{\mathbf{H}}_k + \boldsymbol{\Delta}_k \right)^H \mathbf{v} \right\|^2 = \mathbf{v}^H \left(\hat{\mathbf{H}}_k + \boldsymbol{\Delta}_k \right) \left(\hat{\mathbf{H}}_k + \boldsymbol{\Delta}_k \right)^H \mathbf{v}$$

$$\approx \mathbf{v}^H \hat{\mathbf{H}}_k \hat{\mathbf{H}}_k^H \mathbf{v} + 2\,\text{Re}\left\{ tr\left(\mathbf{v}^H \boldsymbol{\Delta}_k \hat{\mathbf{H}}_k^H \mathbf{v} \right) \right\}, \tag{3.11}$$

where the term $\mathbf{v}^H \boldsymbol{\Delta}_k \boldsymbol{\Delta}_k^H \mathbf{v}$ is negligible because it is much smaller than the other terms. Based on Lemma 3.1, the minimum of term $\text{Re}\left\{ tr\left(\mathbf{v}^H \boldsymbol{\Delta}_k \hat{\mathbf{H}}_k^H \mathbf{v} \right) \right\}$ is given by

$$\text{Re}\left\{\text{tr}\left(\mathbf{v}^H\boldsymbol{\Delta}_k\hat{\mathbf{H}}_k^H\mathbf{v}\right)\right\} = \text{Re}\left\{\text{tr}\left(\boldsymbol{\Delta}_k\hat{\mathbf{H}}_k^H\mathbf{v}\mathbf{v}^H\right)\right\}$$

$$\geq -\varepsilon_k\left\|\hat{\mathbf{H}}_k^H\mathbf{v}\mathbf{v}^H\right\|_F \tag{3.12}$$

To further solve the convexity of constraint (3.9b), we let $\mathbf{V} = \mathbf{v}\mathbf{v}^H$. Thus, constraint (3.9b) becomes convex as follows:

$$\|\mathbf{W}_k\|_F^2 \leq \theta_k\left(\text{tr}(\hat{\mathbf{H}}_k^H\mathbf{V}\hat{\mathbf{H}}_k) - 2\varepsilon_k\left\|\hat{\mathbf{H}}_k^H\mathbf{V}\right\|_F\right). \tag{3.13}$$

Next, we introduce some helpful lemmas to bound the uncertainty of the constraint (3.9c).

Lemma 3.2 (Schur's complement, [18]) *Let* $\mathbf{Y} = \begin{bmatrix} \mathbf{A} & \mathbf{B}^H \\ \mathbf{B} & \mathbf{C} \end{bmatrix}$ *be a Hermitian matrix. Then,* $\mathbf{Y} \succeq \mathbf{0}$ *holds true if and only if* $\mathbf{A} - \mathbf{B}^H\mathbf{C}^{-1}\mathbf{B} \succeq \mathbf{0}$ *with assuming* $\mathbf{C}$ *is invertible, or* $\mathbf{C} - \mathbf{B}^H\mathbf{A}^{-1}\mathbf{B} \succeq \mathbf{0}$ *with assuming* $\mathbf{A}$ *is invertible.*

Lemma 3.3 *Consider a function* $\mathbf{F}(\mathbf{X}) = \mathbf{A} - \left(\mathbf{B}^H\mathbf{X}\mathbf{C} + \mathbf{C}^H\mathbf{X}^H\mathbf{B}\right)$, *where Hermitian matrix* $\mathbf{A}_m \in \mathbb{C}^{L \times L}$, *matrix* $\mathbf{B} \in \mathbb{C}^{M \times L}$, *matrix* $\mathbf{C} \in \mathbb{C}^{N \times L}$, *and variable matrix* $\mathbf{X} \in \mathbb{C}^{M \times N}$. *Then*

$$\mathbf{F}(\mathbf{X}) \succeq \mathbf{0}, \ \forall \mathbf{X} : \|\mathbf{X}\|_F \leq \varepsilon$$

holds true if and only if there exits $\lambda \geq 0$, *such that*

$$\begin{bmatrix} \mathbf{A} - \lambda\mathbf{C}^H\mathbf{C} & -\varepsilon\mathbf{B}^H \\ -\varepsilon\mathbf{B} & \lambda\mathbf{I} \end{bmatrix} \succeq \mathbf{0}.$$

Proof Please refer to Appendix B.

Now, we deal with the constraint (3.9c), the left of which can be transformed as

$$\left\|\mathbf{Z}^H\left(\hat{\mathbf{H}}_k+\boldsymbol{\Delta}_k\right)\mathbf{W}_k - \mathbf{I}\right\|_F^2$$

$$= \left\|\text{vec}\left(\mathbf{Z}^H\left(\hat{\mathbf{H}}_k+\boldsymbol{\Delta}_k\right)\mathbf{W}_k - \mathbf{I}\right)\right\|^2$$

$$= \left\|\text{vec}\left(\mathbf{Z}^H\hat{\mathbf{H}}_k\mathbf{W}_k - \mathbf{I}\right) + \text{vec}\left(\mathbf{Z}^H\boldsymbol{\Delta}_k\mathbf{W}_k\right)\right\|^2,$$

$$= \|\mathbf{w}_k + \mathbf{D}_k\boldsymbol{\delta}_k\|^2, \tag{3.14}$$

where $\mathbf{w}_k \triangleq \text{vec}\left(\mathbf{Z}^H\hat{\mathbf{H}}_k\mathbf{W}_k - \mathbf{I}\right)$, $\mathbf{D}_k \triangleq \mathbf{W}_k^T \otimes \mathbf{Z}^H$ and $\boldsymbol{\delta}_k = \text{vec}(\boldsymbol{\Delta}_k)$. According to Lemma 3.2, $\|\mathbf{w}_k + \mathbf{D}_k\boldsymbol{\delta}_k\|^2 \leq \eta_k$ can be rewritten as

$$\begin{bmatrix} \eta_k & (\mathbf{w}_k + \mathbf{D}_k\boldsymbol{\delta}_k)^H \\ \mathbf{w}_k + \mathbf{D}_k\boldsymbol{\delta}_k & \mathbf{I} \end{bmatrix} \succeq \mathbf{0}, \ \forall\boldsymbol{\delta}_k : \|\boldsymbol{\delta}_k\|_2 \leq \varepsilon_k. \tag{3.15}$$

Then, we let

$$\mathbf{A}_k \triangleq \begin{bmatrix} \eta_k & \mathbf{w}_k^H \\ \mathbf{w}_k & \mathbf{I} \end{bmatrix}, \mathbf{B}_k \triangleq \begin{bmatrix} \mathbf{0} & -\mathbf{D}_k^H \end{bmatrix}, \mathbf{c} \triangleq \begin{bmatrix} 1 & \mathbf{0} \end{bmatrix}. \tag{3.16}$$

In this case, Eq. (3.15) can be further expressed as

$$\mathbf{F}_k (\delta_k) = \mathbf{A}_k - \left(\mathbf{B}_k^H \delta_k \mathbf{c} + \mathbf{c}^H \delta_k^H \mathbf{B}_k \right) \succeq \mathbf{0}. \tag{3.17}$$

Applying Lemma 3.3, the constraint (3.9c) can be reformulated as

$$\begin{bmatrix} \eta_k - \tau_k & \mathbf{w}_k^H & \mathbf{0} \\ \mathbf{w}_k & \mathbf{I} & \varepsilon_k \mathbf{D}_k \\ \mathbf{0} & \varepsilon_k \mathbf{D}_k^H & \tau_k \mathbf{I} \end{bmatrix} \geq \mathbf{0}, \ \exists \tau_k \geq 0. \tag{3.18}$$

Based on the transformed constraint (3.9b) and (3.9c), problem (3.9) can be transformed as

$$\min_{\mathbf{V}, \mathbf{Z}, \{\mathbf{W}_k\}, \{\eta_k\}, \{\tau_k\}} \sum_{k=1}^{K} \eta_k + \sigma_n^2 \|\mathbf{Z}\|_F^2 \tag{3.19a}$$

$$\text{s.t.} \ (5.17d), (4.15), (4.20).$$

$$\tau_k \geq 0, \tag{3.19b}$$

$$\text{tr}(\mathbf{V}) \leq P_{\max}, \tag{3.19c}$$

$$\mathbf{V} \succeq \mathbf{0}, \tag{3.19d}$$

$$\text{Rank}(\mathbf{V}) = 1. \tag{3.19e}$$

Due to coupled variables $(\mathbf{V}, \mathbf{Z}, \{\mathbf{W}_k\})$ in (3.19), it is still an untractable problem. Notice that if $\mathbf{Z}$ is fixed, it becomes a classical convex semi-definite programming (SDP) problem for the variables $\{\mathbf{W}_k\}$ and $\mathbf{V}$ with applying a general semi-definite relaxation (SDR) technique, i.e., dropping the rank-one constraint on $\mathbf{V}$. On the other hand, if $\{\mathbf{W}_k\}$ and $\mathbf{V}$ are fixed, it is also a convex SDP problem for $\mathbf{Z}$. Thus, one can alternatively optimize variables by fixed other variables to effectively solve this problem with the interior-point method [18]. In particular, we utilize an off-the-shelf convex optimization toolbox CVX [19] to solve problem (3.19) at a computational complexity order of at most $\mathcal{O}(K(N_b N_u + L^2)^{3.5})$ for each iteration [20]. For the obtained energy beam, we have the following proposition.

Proposition 3.1 *The optimal solution $\mathbf{V}^*$ to the problem (3.19) always satisfies the rank-one condition $Rank(\mathbf{V}^*) = 1$.*

Proof Please refer to Appendix C.

Therefore, once the problem converges, we can get the unique solution $\mathbf{v}^*$ to the original problem (3.8) by eigenvalue decomposition (EVD) on $\mathbf{V}^*$, which is given by

$$\mathbf{v}^* = \sqrt{\lambda_{\max}(\mathbf{V}^*)}\boldsymbol{\xi}_{\max}, \tag{3.20}$$

where $\lambda_{\max}(\mathbf{V}^*)$ is the maximum eigenvalue of $\mathbf{V}^*$ and $\boldsymbol{\xi}_{\max}$ is the unit eigenvector related to $\lambda_{\max}(\mathbf{V}^*)$. In summary, the robust design for computation-centric B5G cellular IoT can be described as Algorithm 3.1.

Algorithm 3.1 Robust Design for computation-centric B5G Cellular IoT

Input: $N_b, N_u, L, K, \sigma_n^2, P_{\max}, \theta_k, \varepsilon_k, \forall k = 1, ..., K$
Output: $\mathbf{v}, \mathbf{Z}, \mathbf{W}_k$.
1: **Initialize** $\mathbf{Z}, \mathbf{V} = \mathbf{v}\mathbf{v}^H$;
2: **repeat**
3:   find $\{\mathbf{V}^*, \mathbf{W}_k^*\}$ for solving the problem (3.19) by CVX with fixed $\mathbf{Z}$;
4:   find $\mathbf{Z}^*$ for solving the problem (3.19) by CVX with fixed $\{\mathbf{W}_k^*\}$;
5: **until** convergence
6: obtain $\mathbf{v}^*$ by EVD on $\mathbf{V}^*$ according to (3.20).

3.4 Numerical Results

In this section, we provide some simulation results to validate the robustness and effectiveness of the proposed Algorithm 3.1 for the computation-centric B5G cellular IoT. Unless extra specification, the simulation parameters are set as in Table 3.2. All IoT UEs are distributed within a range of 10 to 100 meters from the BS independently, and a general urban channel pass-loss model with $\mathrm{PL}_{\mathrm{dB}} = 10\log_{10}(d^2) + 10$ is adopted [21]. For the sake of observation, we refer to normalized computation error MSE/K as the performance metric, and use $\mathrm{SNR} = 10\log_{10}(P_{\max}/\sigma_n^2)$ to denote the transmit SNR (in dB).

Table 3.2 Simulation parameters

Parameters	Values
BS parameter	$N_b = 32$
IoT UEs parameters	$K = 24, N_u = L = 2$
Distance (d) between IoT UEs and the BS	$10 \sim 100$ (m)
Power conversion efficiency	$\theta_k = \theta = 0.5$
Noise powers	$\sigma_n^2 = 1$
Radius of the channel uncertainty region	$\varepsilon_k = \varepsilon = 0.01$

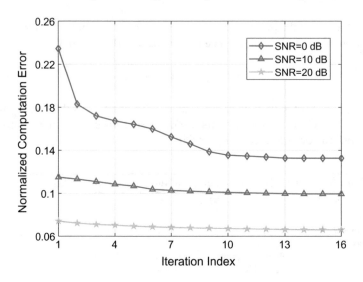

Fig. 3.2 Convergence behavior of the proposed Algorithm 3.1

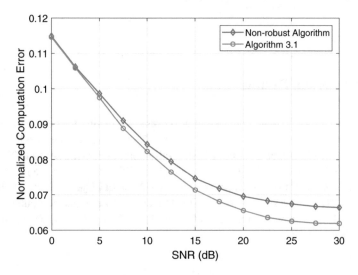

Fig. 3.3 Normalized Computation Error versus SNR for different algorithms

First, we show the convergence behaviors of the proposed Algorithm 3.1 under different SNR values in Fig. 3.2. It is observed that Algorithm 3.1 converges after a few number of iterations under different SNR values, which means the complexity cost is affordable for practical implementation. In addition, Algorithm 3.1 has a steady convergence at higher SNR, while it requires more iterations at lower SNR. This is because the received data at the BS contains huge interference in the low transmit SNR region.

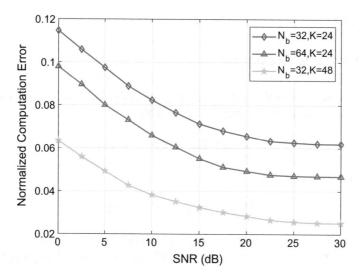

Fig. 3.4 Normalized Computation Error versus SNR for different number of IoT UEs and BS antennas

Next, in Fig. 3.3, we illustrate the performance of the proposed Algorithm 3.1 over the non-robust one that regards the estimated channel $\{\hat{\mathbf{H}}_k\}$ as the real channel. It is seen that Algorithm 3.1 taking into account the CSI uncertainty in the design always outperforms the non-robust one in the whole region, and the gap between them enlarges as the SNR increases, which demonstrates the robustness and effectiveness of Algorithm 3.1.

Figure 3.4 reveals the impacts of numbers of BS antennas N_b and IoT UEs K on the performance. First, it is obvious that the normalized computation error decreases as the number of BS antennas N_b increases, since the BS can provide more array gains for enhancing the performance. Then, as the number of IoT UEs K increases, the normalized computation error also decreases. This is because the combined received signal will be accordingly enhanced with the number of IoT UEs. Thus, Algorithm 3.1 can achieve more performance gains when the number of accessed IoT UEs is large, which exactly means Algorithm 3.1 is quite suitable to the computation-centric B5G cellular IoT network.

3.5 Conclusion

This chapter has designed a computation-centric B5G cellular IoT combing with MIMO-AirComp and WPT. For realizing effective convergence of energy supply and data aggregation under practical conditions, a robust beamforming design algorithm has been provided. Extensive simulations verified that the proposed algorithm can improve the overall performance of B5G cellular IoT.

Appendix

A: The Proof of Lemma 3.2

It is known that $-|\text{tr}\,(\mathbf{XY})| \leq \text{Re}\,\{\text{tr}\,(\mathbf{XY})\} \leq |\text{tr}\,(\mathbf{XY})|$. Then, according to the definition of trace and norm for matrix [22], we have

$$|\text{tr}\,(\mathbf{XY})| \leq \|\mathbf{X}\|_F \|\mathbf{Y}\|_F \leq \delta \|\mathbf{Y}\|_F. \tag{3.21}$$

As such, we can obtain

$$-\delta \|\mathbf{Y}\|_F \leq \text{Re}\,\{\text{tr}\,(\mathbf{XY})\} \leq \delta \|\mathbf{Y}\|_F. \tag{3.22}$$

Specifically, the lower bound and upper bound of $\text{Re}\,\{\text{tr}\,(\mathbf{XY})\}$ are $-\delta \|\mathbf{Y}\|_F$ with $\mathbf{X} = -\delta \mathbf{Y}/\|\mathbf{Y}\|_F$ and $\delta \|\mathbf{Y}\|_F$ with $\mathbf{X} = \delta \mathbf{Y}/\|\mathbf{Y}\|_F$, respectively. The proof is completed.

B: The Proof of Lemma 3.3

Based on the definition of semi-positive matrix, we have

$$\mathbf{F}\,(\mathbf{X}) \succeq \mathbf{0}, \; \forall \mathbf{X} : \|\mathbf{X}\|_F \leq \varepsilon$$

if and only if for $\forall \mathbf{d} \neq \mathbf{0}$

$$\mathbf{d}^H \mathbf{A} \mathbf{d} - \mathbf{d}^H \left(\mathbf{B}^H \mathbf{X} \mathbf{C} + \mathbf{C}^H \mathbf{X}^H \mathbf{B}\right) \mathbf{d} \geq 0. \tag{3.23}$$

According to Lemma 3.2, (3.23) is equivalent to

$$\mathbf{d}^H \mathbf{A} \mathbf{d} \geq \max_{\|\mathbf{X}\|_F \leq \varepsilon} \mathbf{d}^H \left(\mathbf{B}^H \mathbf{X} \mathbf{C} + \mathbf{C}^H \mathbf{X}^H \mathbf{B}\right) \mathbf{d}$$
$$= 2\varepsilon \|\mathbf{C}\mathbf{d}\| \|\mathbf{B}\mathbf{d}\|. \tag{3.24}$$

Applying the *Cauchy-Schwarz inequality*, (3.24) can be further transformed as

$$\mathbf{d}^H \mathbf{A} \mathbf{d} - 2\varepsilon \mathbf{e}^H \mathbf{C} \mathbf{d} \geq 0, \forall \mathbf{e} : \|\mathbf{e}\| \leq \|\mathbf{B}\mathbf{d}\|. \tag{3.25}$$

To further deal with (3.25), we need to introduce the following lemma:

Lemma 3.4 (S-procedure, [18]) *Let us consider a function* $\mathbf{f}_m\,(\mathbf{x}) = \mathbf{x}^H \mathbf{A}_m \mathbf{x} + 2\,\text{Re}\,\{\mathbf{b}_m^H \mathbf{x}\} + \mathbf{c}_m, m \in \{1, 2\}, \mathbf{x} \in \mathbb{C}^{N \times 1}$, where $\mathbf{A}_m \in \mathbb{C}^{N \times N}, \mathbf{b}_m \in \mathbb{C}^{N \times 1}$ and $\mathbf{c}_m \in \mathbb{C}^{N \times 1}$ are known. The derivation $\mathbf{f}_1\,(\mathbf{x}) \leq 0 \Rightarrow \mathbf{f}_2\,(\mathbf{x}) \leq 0$ holds true if and only if there exists a $\tau \geq 0$, such that

$$\tau \begin{bmatrix} \mathbf{A}_1 & \mathbf{b}_1 \\ \mathbf{b}_1^H & \mathbf{c}_1 \end{bmatrix} - \begin{bmatrix} \mathbf{A}_2 & \mathbf{b}_2 \\ \mathbf{b}_2^H & \mathbf{c}_2 \end{bmatrix} \succeq \mathbf{0}. \tag{3.26}$$

By exploiting Lemma 3.4 with $\mathbf{ee}^H - \mathbf{d}^H \mathbf{B}^H \mathbf{B} \mathbf{d} \leq 0$, (3.25) is satisfied if and only if there exists a $\lambda \geq 0$ such that

$$\begin{bmatrix} \mathbf{A} - \lambda \mathbf{C}^H \mathbf{C} & -\varepsilon \mathbf{B}^H \\ -\varepsilon \mathbf{B} & \lambda \mathbf{I} \end{bmatrix} \succeq \mathbf{0}.$$

The proof is completed.

C: The Proof of Proposition 3.1

The lagrangian function of problem (3.19) with respect to $\mathbf{V}$ can be expressed as

$$\mathscr{L}(\mathbf{V}) = \alpha \left(\operatorname{tr}(\mathbf{V}) - P_{\max} \right) - \boldsymbol{\Upsilon} \mathbf{V} - \sum_{k=1}^{K} \beta_k \tag{3.27}$$
$$\times \left[\theta_k \operatorname{tr}\left(\hat{\mathbf{H}}_k^H \mathbf{V} \hat{\mathbf{H}}_k \right) - 2\theta_k \varepsilon_k \left\| \hat{\mathbf{H}}_k^H \mathbf{V} \right\|_F - \|\mathbf{W}_k\|_F^2 \right],$$

where $\alpha \geq 0$, $\beta_k \geq 0$ and $\boldsymbol{\Upsilon} \succeq 0$ are Lagrange multipliers of constraint (3.19c), (3.13) and (3.19d), respectively. Satisfied with Slater's condition, we reveal the structure of the optimal $\mathbf{V}^*$ by exploiting Karush-Kuhn-Tucker (KKT) conditions:

$$\Theta(\mathbf{V}^*) = \theta_k \operatorname{tr}\left(\hat{\mathbf{H}}_k^H \mathbf{V}^* \hat{\mathbf{H}}_k \right) - 2\theta_k \varepsilon_k \left\| \hat{\mathbf{H}}_k^H \mathbf{V}^* \right\|_F - \|\mathbf{W}_k\|_F^2 = 0, \tag{3.28a}$$

$$\operatorname{tr}(\mathbf{V}^*) - P_{\max} = 0 \tag{3.28b}$$

$$\boldsymbol{\Upsilon}^* \mathbf{V}^* = \mathbf{0}, \tag{3.28c}$$

$$\nabla_{\mathbf{V}^*} \mathscr{L}(\mathbf{V}^*) = \alpha^* \mathbf{I} - \boldsymbol{\Upsilon}^* - \beta_k^* \nabla_{\mathbf{V}^*} \Theta(\mathbf{V}^*) = 0, \tag{3.28d}$$

$$\alpha^* \geq 0, \beta_k^* \geq 0, \boldsymbol{\Upsilon}^* \succeq \mathbf{0}. \tag{3.28e}$$

From (3.28b), due to $P_{\max} > 0$, we can obtain that $\mathbf{V}^* \neq \mathbf{0}$, namely,

$$\operatorname{Rank}(\mathbf{V}^*) \geq 1. \tag{3.29}$$

Based on the *Sylvesters rank inequality* [23], we can deduce that $\operatorname{Rank}(\boldsymbol{\Upsilon}^*) + \operatorname{Rank}(\mathbf{V}^*) \leq N_b$ due to $\boldsymbol{\Upsilon}^* \mathbf{V}^* = \mathbf{0}$ from (3.28c). That is,

$$\operatorname{Rank}(\boldsymbol{\Upsilon}^*) \leq N_b - 1. \tag{3.30}$$

Then, in term of (3.28d), we propose the following lemma:

Lemma 3.5 *For two matrices $\mathbf{X}$ and $\mathbf{Y}$ with the same size, it is always true that $Rank(\mathbf{X} + \mathbf{Y}) \leq Rank(\mathbf{X}) + Rank(\mathbf{Y})$.*

Proof

$$\text{Rank}(\mathbf{X} + \mathbf{Y}) = \text{Rank} \begin{bmatrix} \mathbf{X} + \mathbf{Y} \\ \mathbf{0} \end{bmatrix} \leq \text{Rank} \begin{bmatrix} \mathbf{X} + \mathbf{Y} \\ \mathbf{Y} \end{bmatrix}$$
$$= \text{Rank} \begin{bmatrix} \mathbf{X} \\ \mathbf{Y} \end{bmatrix} \leq \text{Rank}(\mathbf{X}) + \text{Rank}(\mathbf{Y}). \tag{3.31}$$

According to Lemma 3.5, we have

$$\text{Rank}(\boldsymbol{\Upsilon}^*) + \text{Rank}\left(\beta_k^* \nabla_{\mathbf{V}^*} \Theta \left(\mathbf{V}^*\right)\right) \geq \text{Rank}\left(\alpha^* \mathbf{I}\right), \tag{3.32}$$

where $\nabla_{\mathbf{V}^*} \Theta \left(\mathbf{V}^*\right) \neq \mathbf{0}$ due to $\mathbf{V}^* \neq \mathbf{0}$, i.e., $\text{Rank}\left(\beta_k \nabla_{\mathbf{V}^*} \Theta \left(\mathbf{V}^*\right)\right) \geq 1$. Thus, (3.32) can be further expressed as

$$\text{Rank}(\boldsymbol{\Upsilon}^*) \geq N_b - 1. \tag{3.33}$$

Combining (3.30) and (3.33), it is easily seen that $\text{Rank}(\boldsymbol{\Upsilon}^*) = N_b - 1$. Finally, According to the *Sylvesters rank inequality* [23], we have

$$\text{Rank}\left(\boldsymbol{\Upsilon}^* \mathbf{V}^*\right) \geq \text{Rank}(\boldsymbol{\Upsilon}^*) + \text{Rank}(\mathbf{V}^*) - N_b, \tag{3.34}$$

where $\text{Rank}\left(\boldsymbol{\Upsilon}^* \mathbf{V}^*\right) = 0$ from (3.28c) and $\text{Rank}(\boldsymbol{\Upsilon}^*) = N_b - 1$. Thus, we can obtain that

$$\text{Rank}(\mathbf{V}^*) \leq 1. \tag{3.35}$$

By (3.29) and (3.35), we can come to a conclusion that $\text{Rank}(\mathbf{V}^*) = 1$, which means the SDR processing for $\mathbf{V} = \mathbf{v}\mathbf{v}^H$ in the problem (3.19) is tight. The proof is completed.

References

1. Chen X, Ng DWK, Yu W, Larsson EG, Al-Dhahir N, Schober R (2020) Massive access for 5G and beyond. IEEE J Sel Area Commun PP(99):1–24
2. Chen X (2019) Massive access for cellular internet of things theory and technique. Springer, Germany
3. Zihan E, Choi KW, Kim DI (2015) Distributed random access scheme for collision avoidance in cellular device-to-device communication. IEEE Trans Wirel Commun 14(7):3571–3585
4. Abari O, Rahul H, Katabi D (2016) Over-the-Air function computation in sensor networks. http://arxiv.org/pdf/1612.02307.pdf
5. Nazer B, Gastpar M (2007) Computation over multiple-access channels. IEEE Trans Inf Theory 53(10):3498–3516

6. Goldenbaum M, Boche H, Staczak S (2015) Nomographic functions: efficient computation in clustered gaussian sensor networks. IEEE Trans Wirel Commun 14(4):2093–2105
7. Goldenbaum M, Boche H, Staczak S (2013) Harnessing interference for analog function computation in wireless sensor networks. IEEE Trans Signal Process 61(20):4893–4906
8. Qi Q, Chen X, Lei L, Zhong C, Zhang Z (2019) Robust convergence of energy and computation for B5G cellular internet of things. In: Proceeding of IEEE GLOBECOM, Waikoloa, USA, pp 1–6
9. Zhu G, Huang K (2019) MIMO over-the-air computation for high-mobility multimodal sensing. IEEE Internet Things J 6(4):6089–6103
10. Wen D, Zhu G, Huang K (2019) Reduced-dimension design of MIMO over-the-air computing for data aggregation in clustered IoT networks. IEEE Trans Wirel Commun 18(11):5255–5268
11. Chen X, Zhang Z, Chen H-H, Zhang H (2015) Enhancing wireless information and power transfer by exploiting multiantenna techniques. IEEE Commun Mag 53(4):133–141
12. Qi Q, Chen X (2019) Wireless powered massive access for cellular internet of things with imperfect SIC and nonlinear EH. IEEE Internet Things J 6(2):3110–3120
13. Qi Q, Chen X, Lei L, Zhong C, Zhang Z (2019) Outage-constrained robust design for sustainable B5G cellular internet of things. IEEE Trans Wirel Commun 18(12):5780–5790
14. Chen X, Yuen C, Zhang Z (2014) Wireless energy and information transfer tradeoff for limited feedback multi-antenna systems with energy beamforming. IEEE Trans Veh Technol 63(1):407–412
15. Chen X, Wang X, Chen X (2013) Energy-efficient optimization for wireless information and power transfer in large-scale MIMO systems employing energy beamforming. IEEE Wirel Commun Lett 2(6):667–670
16. Qi Q, Chen X, Ng DWK (2020) Robust beamforming for NOMA-based cellular massive IoT with SWIPT. IEEE Trans Signal Process PP(99):1–15 (2020)
17. Wang J, Palomar DP (2009) Worst-case robust MIMO transmission with imperfect channel knowledge. IEEE Trans Signal Process 57(8):3086–3100
18. Boyd S, Vandenberghe L (2004) Convex optimization. Cambridge University, Cambridge, UK
19. Grant M, Boyd S CVX: matlab software for disciplined convex programming. http://cvxr.com/cvx
20. Ben-Tal A, Nemirovski A (2001) Lectures on modern convex optimization: Analysis, algorithms, and engineering applications. SIAM, MPS-SIAM Series on Optimization, Philadelphia, PA, USA
21. Hata M (1980) Empirical formula for propagation loss in land mobile radio services. IEEE Trans Veh Technol 29(3):317–325
22. Horn A, Charles R (1985) Matrix analysis. Cambridge University, Cambridge, UK
23. Matsaglia G, Styan GPH (1974) Equalities and inequalities for ranks of matrices. Linear Multilinear Algebra 2:269–292

Chapter 4
Convergence of Communication and Computation in B5G Cellular Internet of Things

Abstract In this chapter, we investigate the issue of convergence of computation and communication in B5G cellular IoT with multiple tasks. According to the characteristics of B5G cellular IoT, a comprehensive deign framework integrating computation and communication is put forward for massive IoT. For communication, highly accurate sensing information at IoT devices are sent to the BS through non-orthogonal communication over wireless multiple access channels. Meanwhile, for computation, AirComp is adopted to substantially reduce latency of massive data aggregation via exploiting the superposition property of wireless multiple-access channels. To achieve an effective integration of computation and communication under practical but adverse conditions, a robust algorithm is proposed by jointly optimizing transmit power and receive beamforming with the goal of minimizing the computation error of computation signals while guaranteeing the quality of communication signals. Finally, extensive simulations validate the robustness and effectiveness of the proposed algorithm for B5G cellular IoT.

4.1 Introduction

Advanced sensors enable IoT devices to capture vast amounts of data generated by human or surrounding environment [1]. As a result, the measured mass data needs to be transmitted to the base station (BS) for further decoding, computation or analysis. Hence, computation and communication are two fundamental tasks for B5G cellular IoT [2]. However, it is challenging to complete the two tasks with limited wireless resources. On the one hand, for communication, traditional orthogonal multiple access (OMA) schemes cannot support high-capacity transmission of a massive number of IoT devices due to limited radio spectrum. To solve this challenge, non-orthogonal multiple access (NOMA) is applied into cellular IoT to realize massive access [3, 4]. Nevertheless, massive NOMA leads to severe co-channel information, which limits the performance of massive access. Commonly, spatial beamforming is utilized to combat co-channel interference [5–7]. Especially in B5G cellular IoT, the BS equipped with a large-scale antenna array has ultra-high spatial degrees of freedom to mitigate co-channel interference [8, 9].

On the other hand, for computation, it is intuitive that the conventional way of *transmit-then-compute* no longer suits the demands of ultra low latency and high spectrum utilization for massive IoT [10]. To this end, *over-the-air computation* (AirComp), as a feasible solution exploiting the superposition property of wireless multiple-access channel was proposed [11] to compute a class of *nomographic functions* of distributed data from IoT devices via concurrent transmission. Unlike classical wireless communication systems where simultaneous transmissions lead to harmful interference, the accuracy of computation enabled by AirComp is improved with the increment of simultaneous IoT devices [12]. In particular, AirComp can significantly decrease the data aggregation latency via a factor equal to the number of IoT devices by making use of interference from concurrent transmission for functional computation. In B5G cellular IoT, AirComp can also combine with spatial beamforming to reduce the distortion of computation results. In [13], a beamforming optimization for AirComp in sensors network was well analyzed and designed. In [14], a joint beamforming design of energy supply and data aggregation was proposed for wirelessly powered AirComp systems.

In general, it is expected to realize the two tasks simultaneously by exploiting the open nature of wireless channels. However, it is not a straightforward issue to effectively integrate computation and communication in B5G cellular IoT, especially under practical condition that channel state information (CSI) can not be perfectly obtained. Moreover, there is no relevant work involving beamforming design both for communication and computation in massive IoT. In this paper, we design a general B5G cellular IoT network integrating communication and computation, where transmit power and receive beamforming are jointly optimized to coordinate the performance of the two tasks with limited wireless resources. The contributions of this paper are two-fold:

1. We propose a comprehensive design framework for the integration of computation and communication. The originally harmful interference caused by simultaneous transmission is exploited to enhance integration performance.
2. We analyze the impacts of transmit power and receive beamforming on the performance of both communication and computation, and then provide a robust algorithm to balance the performance of computation and communication under practical conditions.

The rest of this chapter is organized as follows. Section 4.2 offers a integration framework of computation and communication in B5G cellular IoT networks. Section 4.3 focuses on the design of a robust algorithm for the proposed framework. Section 4.4 presents several simulation results to validate the effectiveness and robustness of the proposed algorithm. Finally, Sect. 4.5 summarizes the chapter.

Notations: We use bold upper (lower) letters to denote matrices (column vectors), $(\cdot)^H$ to denote conjugate transpose, $\| \cdot \|$ to denote the L_2-norm of a vector, $| \cdot |$ to denote the absolute value, $\text{Re}\{\cdot\}$ to denote the real parts of a complex number, $\mathbb{E}\{\cdot\}$ to denote the expectation value, $\text{tr}(\cdot)$ to denote trace of a matrix, and $\text{Rank}(\cdot)$ to denote rank of a matrix.

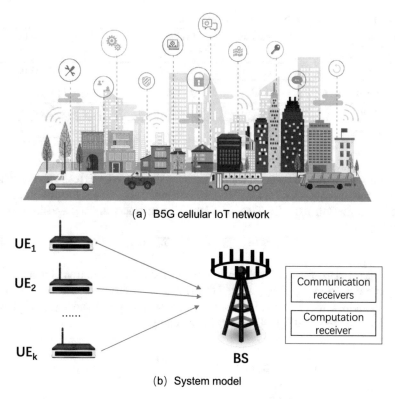

(a) B5G cellular IoT network

(b) System model

Fig. 4.1 A framework integrating computation and communication for B5G cellular IoT

4.2 System Model

Let us consider a B5G cellular IoT network, where a BS equipped with N antennas serves K single-antenna IoT user equipments (UEs), c.f., Fig. 4.1. Each IoT UE has two independent signals, one for computation, and the other for communication. The two signals are first superposition coded, and then sent to the BS over the uplink channel. On the one hand, by exploiting the superposition property of wireless channel, the BS receives the computation results directly via concurrent data transmission without recovering individual data, and then utilizes a computation receiver to obtain the targeted function signal. To realize different computation functions, the computation signal enabled by Aircomp needs to be pre-processed and post-processed at the IoT UEs and the BS, respectively. Some typical processing functions are given in Table 4.1. On the other hand, the BS decodes the communication signals of each IoT UE through communication receivers.

Without loss of generality, it is assumed that the kth IoT UE records measured data including a parameter to be computed and a parameter to be communicated from the environment or human in each time slot. Let $s_{k,1} = g_k (d_k)$ and $s_{k,2}$ denote the

Table 4.1 Some examples of nomographic functions

Functions	Pre-processing	Post-processing	Targeted function
Arithmetic mean	$g_k = d_k$	$f = 1/K$	$q = \frac{1}{K} \sum_{k=1}^{K} d_k$
Weighted sum	$g_k = \vartheta_k d_k$	$f = 1$	$q = \sum_{k=1}^{K} \vartheta_k d_k$
Geometric mean	$g_k = \ln(d_k)$	$f = \exp(\cdot)$	$q = \left(\prod_{k=1}^{K} d_k \right)^{1/K}$
Polynomial	$g_k = \vartheta_k d_k^{\beta_k}$	$f = 1$	$q = \sum_{k=1}^{K} \vartheta_k d_k^{\beta_k}$
Euclidean norm	$g_k = d_k^2$	$f = \sqrt{(\cdot)}$	$q = \sqrt{\sum_{k=1}^{K} d_k^2}$

computation signal and communication signal at the kth IoT UE respectively, where d_k is the measured value of parameter to be computed and $g_k(\cdot)$ is the pre-processing function at the kth IoT UE. Thus, the kth UE constructs the superposition coded signal x_k as

$$x_k = \sqrt{p_{k,1}} s_{k,1} + \sqrt{p_{k,2}} s_{k,2}, \qquad (4.1)$$

where $p_{k,1}$ and $p_{k,2}$ denote the transmit power for the computation signal $s_{k,1}$ and the communication signal $s_{k,2}$ at the kth IoT UE, respectively. For ease of analysis but without loss of generality, we assume that $\mathbb{E}\left\{ s_{k,1} s_{k,1}^H \right\} = 1$ and $\mathbb{E}\left\{ s_{k,2} s_{k,2}^H \right\} = 1$. Therefore, the received signal at the BS is given by

$$
\begin{aligned}
\mathbf{y} &= \sum_{k=1}^{K} \mathbf{h}_k x_k + \mathbf{n} \\
&= \underbrace{\sum_{k=1}^{K} \mathbf{h}_k \sqrt{p_{k,1}} s_{k,1}}_{\text{computation signal}} + \underbrace{\sum_{k=1}^{K} \mathbf{h}_k \sqrt{p_{k,2}} s_{k,2}}_{\text{communication signal}} + \mathbf{n},
\end{aligned}
\qquad (4.2)
$$

where $\mathbf{n}$ is the additive white Gaussian noise (AWGN) vector with the distribution $\mathscr{CN}(\mathbf{0}, \sigma_n^2 \mathbf{I})$, and $\mathbf{h}_k$ denotes the channel vector from the kth UE to the BS. It is reasonably assumed that $\mathbf{h}_k$ remains unchanged during a time slot, but independently fades over time slots. Now, we consider the processing of the computation signal. The BS engages itself in computing the nomographic function [15], such that

$$q = f \left(\sum_{k=1}^{K} s_{k,1} \right), \qquad (4.3)$$

where $f(\cdot)$ represents the post-processing function at the BS. Due to the one-to-one mapping between $s = \sum_{k=1}^{K} s_{k,1}$ and q in (4.3), we take an accurate s at the BS as the targeted function signal. To minimize the distortion of the targeted function signal caused by channel fading, noise and interference, it is necessary to perform receive beamforming at the BS. Thus, the received signal for computation at the BS is transformed as

$$\hat{s} = \mathbf{z}^H \sum_{k=1}^{K} \left[\mathbf{h}_k \left(\sqrt{p_{k,1}} s_{k,1} + \sqrt{p_{k,2}} s_{k,2} \right) \right] + \mathbf{z}^H \mathbf{n}, \tag{4.4}$$

where $\mathbf{z}$ denotes the N-dimensional receive computation beamforming vector at the BS. Mathematically, the accuracy of computation at the BS can be measured by the mean-square-error (MSE) between s and $\hat{s}$, which is given by

$$\text{MSE}\,(\hat{s}, s) = \mathbb{E}\{(\hat{s} - s)(\hat{s} - s)^H\}. \tag{4.5}$$

Substituting (4.4) into (4.5), the computation distortion can be expressed as the following MSE function in terms of transmit power and receive computation beam:

$$\text{MSE}\,(p_{k,1}, p_{k,2}, \mathbf{z}) = \sum_{k=1}^{K} \left| \mathbf{z}^H \mathbf{h}_k \sqrt{p_{k,1}} - 1 \right|^2 + \sigma_n^2 \|\mathbf{z}\|^2 + \sum_{k=1}^{K} \left| \mathbf{z}^H \mathbf{h}_k \sqrt{p_{k,2}} \right|^2. \tag{4.6}$$

Then, we analyze the processing of the communication signal. The received communication signal sent from the kth UE to the BS is given by

$$y_k' = \mathbf{u}_k^H \mathbf{h}_k \sqrt{p_{k,2}} s_{k,2} + \mathbf{u}_k^H \sum_{i=1,i\neq k}^{K} \mathbf{h}_i \sqrt{p_{i,2}} s_{i,2} + \mathbf{u}_k^H \sum_{i=1}^{K} \mathbf{h}_i \sqrt{p_{i,1}} s_{i,1} + \mathbf{u}_k^H \mathbf{n}, \tag{4.7}$$

where $\mathbf{u}_k$ denotes the receive communication beamforming vector of the communication signal for the kth UE at the BS. As a consequence, the corresponding received signal-to-interference-plus-noise ratio (SINR) at the kth communication receiver can be expressed as

$$\Gamma_k = \frac{\left| \mathbf{u}_k^H \mathbf{h}_k \sqrt{p_{k,2}} \right|^2}{\sum\limits_{i=1,i\neq k}^{K} \left| \mathbf{u}_k^H \mathbf{h}_i \sqrt{p_{i,2}} \right|^2 + \sum\limits_{i=1}^{K} \left| \mathbf{u}_k^H \mathbf{h}_i \sqrt{p_{i,1}} \right|^2 + \sigma_n^2 \|\mathbf{u}_k\|^2}. \tag{4.8}$$

As seen from (4.6) and (4.8), the system performance is jointly affected by transmit power at the IoT UEs, and receive beam at the BS. Hence, it makes sense to optimize the computation and communication parameters according to instantaneous CSI. However, in practical B5G cellular IoT with massive connectivity, it is only able to obtain partial CSI by estimation or feedback. In other words, it is necessary to take channel uncertainty into consideration during the integration of computation and communication, namely robust integration. To characterize the CSI uncertainty, we adopt a commonly-used deterministic model [16]. In particular, the real CSI $\mathbf{h}_k$ related to the kth IoT UE can be modeled as

$$\mathscr{H}_k \triangleq \left\{ \mathbf{h}_k = \hat{\mathbf{h}}_k + \mathbf{e}_k \,\middle|\, \|\mathbf{e}_k\| \leq \varepsilon_k \right\}, \tag{4.9}$$

where $\hat{\mathbf{h}}_k$ is the obtained CSI and $\mathbf{e}_k$ is the channel error vector, whose norm is bounded by a given radius ε_k. In what follows, we improve the integration performance of computation and communication by jointly optimizing transmit power and receive beam in the presence of partial CSI.

4.3 Problem Formulation and Optimization Solution

In this section, we propose a robust design algorithm for the integration of computation and communication with the goal of minimizing the computation error while meeting the requirement of SINR for communication signals. The design can be formulated as the following worst-case optimization problem:

$$\min_{\mathbf{z},\mathbf{u}_k,p_{k,1},p_{k,2},\forall k} \quad \max_{\mathbf{H}_k \in H_k} \mathrm{MSE}\left(p_{k,1}, p_{k,2}, \mathbf{z}\right) \tag{4.10a}$$

$$\mathrm{s.t.} \quad \min_{\mathbf{H}_k \in H_k} \Gamma_k \geq \gamma_k, \tag{4.10b}$$

$$p_{k,1} + p_{k,2} \leq P_{\mathrm{max},k}, \tag{4.10c}$$

where $P_{\mathrm{max},k}$ and γ_k are the maximum transmit power budget and the minimum required SINR of the communication signal from the kth IoT UE, respectively. It is seen that problem (4.10) is non-convex due to the coupled variables, i.e., transmit power $\{p_{k,1}, p_{k,2}\}$ and receive beams $\{\mathbf{z}, \mathbf{u}_k\}$. To solve this issue, we adopt an alternative optimization (AO) algorithm by dividing problem (4.10) into two subproblems, one for transmit power allocation and the other for receive beamforming design. The AO algorithm will stop until the value of the objective function of the original problem approaches a stationary point in the iterations. Firstly, we handle the subproblem of optimizing receive beamforming for given transmit power. By introducing two auxiliary variables α_k and β_k, the subproblem is equivalent to

$$\min_{\mathbf{z},\mathbf{u}_k,\alpha_k,\beta_k,\forall k} \quad \sum_{k=1}^{K} (\alpha_k + \beta_k) + \sigma_n^2 \|\mathbf{z}\|^2 \tag{4.11a}$$

$$\mathrm{s.t.} \quad \frac{1+\gamma_k}{\gamma_k} p_{k,2} \left|\mathbf{u}_k^H \left(\hat{\mathbf{h}}_k + \mathbf{e}_k\right)\right|^2 \geq \sigma_n^2 \|\mathbf{u}_k\|^2 +$$

$$\sum_{i=1}^{K} (p_{i,1} + p_{i,2}) \left|\mathbf{u}_k^H \left(\hat{\mathbf{h}}_i + \mathbf{e}_i\right)\right|^2, \tag{4.11b}$$

$$\left|\mathbf{z}^H \left(\hat{\mathbf{h}}_k + \mathbf{e}_k\right) \sqrt{p_{k,1}} - 1\right|^2 \leq \alpha_k, \tag{4.11c}$$

$$\left|\mathbf{z}^H \left(\hat{\mathbf{h}}_k + \mathbf{e}_k\right) \sqrt{p_{k,2}}\right|^2 \leq \beta_k, \tag{4.11d}$$

$$\alpha_k \geq 0, \tag{4.11e}$$

$$\beta_k \geq 0, \tag{4.11f}$$

$$\forall \mathbf{e}_k : \|\mathbf{e}_k\| \leq \varepsilon_k.$$

We have to address the non-convex constraints (4.11b), (4.11c) and (4.11d), which all involve the channel uncertainty. To handle SINR constraint (4.11b), the following lemma is required:

Lemma 4.1 *If there exists a function $f(\mathbf{p}) = \mathrm{Re}\{\mathbf{p}^H \mathbf{q}\}$ with the domain $\{\mathbf{p} : \|\mathbf{p}\| \leq \delta\}$, then its range is given by*

$$-\delta\|\mathbf{q}\| \leq f(\mathbf{p}) \leq \delta\|\mathbf{q}\|. \tag{4.12}$$

Proof Please refer to Appendix A.

By applying Lemma 4.1, we can obtain the range of the term $\left|\mathbf{u}_k^H \left(\hat{\mathbf{h}}_k + \mathbf{e}_k\right)\right|^2$ in the constraint (4.11b). Specifically, we have

$$
\begin{aligned}
\left|\mathbf{u}_k^H \left(\hat{\mathbf{h}}_k + \mathbf{e}_k\right)\right|^2 &= \mathbf{u}_k^H \left(\hat{\mathbf{h}}_k + \mathbf{e}_k\right) \left(\hat{\mathbf{h}}_k + \mathbf{e}_k\right)^H \mathbf{u}_k \\
&\approx \mathbf{u}_k^H \hat{\mathbf{h}}_k \hat{\mathbf{h}}_k^H \mathbf{u}_k + 2\,\mathrm{Re}\left\{\mathbf{e}_k^H \mathbf{u}_k \mathbf{u}_k^H \hat{\mathbf{h}}_k\right\},
\end{aligned}
\tag{4.13}
$$

where the term $\mathbf{u}_k^H \mathbf{e}_k \mathbf{e}_k^H \mathbf{u}_k$ is negligible because it is much smaller than the other terms. As a result, the range of $\Phi_k = \mathrm{Re}\left\{\mathbf{e}_k^H \mathbf{u}_k \mathbf{u}_k^H \hat{\mathbf{h}}_k\right\}$ is given by

$$-\varepsilon_k \left\|\mathbf{u}_k \mathbf{u}_k^H \hat{\mathbf{h}}_k\right\| \leq \Phi_k \leq \varepsilon_k \left\|\mathbf{u}_k \mathbf{u}_k^H \hat{\mathbf{h}}_k\right\|. \tag{4.14}$$

To further solve the convexity of constraint (4.11b), we define $\mathbf{U}_k = \mathbf{u}_k \mathbf{u}_k^H$. Thus, constraint (4.11b) becomes convex as follows:

$$\frac{1+\gamma_k}{\gamma_k} p_{k,2} \left(\hat{\mathbf{h}}_k^H \mathbf{U}_k \hat{\mathbf{h}}_k - 2\varepsilon_k \left\|\mathbf{U}_k \hat{\mathbf{h}}_k\right\|\right) \geq \sigma_n^2 \mathrm{tr}(\mathbf{U}_k) + \sum_{i=1}^{K} (p_{i,1} + p_{i,2}) \left(\hat{\mathbf{h}}_i^H \mathbf{U}_k \hat{\mathbf{h}}_i + 2\varepsilon_i \left\|\mathbf{U}_k \hat{\mathbf{h}}_i\right\|\right). \tag{4.15}$$

Next, we introduce the following lemmas to bound the uncertainty of the constraint (4.11c).

Lemma 4.2 *(Schur's complement, [17]) Let $\mathbf{M}$ be a Hermitian matrix given by $\mathbf{M} = \begin{bmatrix} \mathbf{A} & \mathbf{B}^H \\ \mathbf{B} & \mathbf{C} \end{bmatrix}$. Then, $\mathbf{M}$ is semi-positive, i.e., $\mathbf{M} \succeq 0$, if and only if $\mathbf{A} - \mathbf{B}^H \mathbf{C}^{-1} \mathbf{B} \succeq 0$ with assuming $\mathbf{C}$ is invertible, or $\mathbf{C} - \mathbf{B}^H \mathbf{A}^{-1} \mathbf{B} \succeq 0$ with assuming $\mathbf{A}$ is invertible.*

Lemma 4.3 *Let us define a matrix function $\mathbf{F}(\mathbf{x}) = \mathbf{A} - \left(\mathbf{B}^H \mathbf{x} \mathbf{c}^H + \mathbf{c} \mathbf{x}^H \mathbf{B}\right)$, where $\mathbf{B} \in \mathbb{C}^{m \times n}$, $\mathbf{x} \in \mathbb{C}^{m \times 1}$, $\mathbf{c} \in \mathbb{C}^{n \times 1}$, and $\mathbf{A} \in \mathbb{C}^{n \times n}$ is a Hermitian matrix. Then*

$$\mathbf{F}(\mathbf{x}) \succeq 0, \quad \forall \mathbf{x} : \|\mathbf{x}\| \leq \varepsilon$$

holds true if and only if there exits $\lambda \geq 0$, such that

$$\begin{bmatrix} \mathbf{A} - \lambda \mathbf{cc}^H & -\varepsilon \mathbf{B}^H \\ -\varepsilon \mathbf{B} & \lambda \mathbf{I} \end{bmatrix} \succeq \mathbf{0}.$$

Proof Please refer to Appendix B.

Now, we deal with the constraint (4.11c), i.e.,

$$\left| \mathbf{z}^H \left(\hat{\mathbf{h}}_k + \mathbf{e}_k \right) \sqrt{p_{k,1}} - 1 \right|^2 = \left| \sqrt{p_{k,1}} \mathbf{z}^H \hat{\mathbf{h}}_k - 1 + \sqrt{p_{k,1}} \mathbf{z}^H \mathbf{e}_k \right|^2$$

$$\leq \alpha_k. \tag{4.16}$$

Based on Lemma 4.2, it can be rewritten as

$$\begin{bmatrix} \alpha_k & \eta_k + \mathbf{w}_k^H \mathbf{e}_k \\ \eta_k + \mathbf{e}_k^H \mathbf{w}_k & 1 \end{bmatrix} \succeq \mathbf{0}, \ \forall \mathbf{e}_k : \|\mathbf{e}_k\| \leq \varepsilon_k. \tag{4.17}$$

where $\eta_k = \sqrt{p_{k,1}} \mathbf{z}^H \hat{\mathbf{h}}_k - 1$, and $\mathbf{w}_k = \sqrt{p_{k,1}} \mathbf{z}$. According to Lemma 4.3, and let

$$\mathbf{A}_k \triangleq \begin{bmatrix} \alpha_k & \eta_k \\ \eta_k & 1 \end{bmatrix}, \mathbf{B}_k \triangleq \begin{bmatrix} \mathbf{0} & -\mathbf{w}_k \end{bmatrix}, \mathbf{c} \triangleq \begin{bmatrix} 1 & 0 \end{bmatrix}^T, \tag{4.18}$$

Equation (4.17) can be further transformed as

$$\mathbf{F}_k(\mathbf{e}_k) = \mathbf{A}_k - \left(\mathbf{B}_k^H \mathbf{e}_k \mathbf{c}^H + \mathbf{c} \mathbf{e}_k^H \mathbf{B}_k \right) \succeq \mathbf{0}, \ \forall \mathbf{e}_k : \|\mathbf{e}_k\| \leq \varepsilon_k. \tag{4.19}$$

Hence, the constraint (4.11c) can be reformulated as

$$\begin{bmatrix} \alpha_k - \phi_k & \mathbf{z}^H \hat{\mathbf{h}}_k \sqrt{p_{k,1}} - 1 & \mathbf{0} \\ \mathbf{z}^H \hat{\mathbf{h}}_k \sqrt{p_{k,1}} - 1 & 1 & \varepsilon_k \sqrt{p_{k,1}} \mathbf{z}^H \\ \mathbf{0} & \varepsilon_k \sqrt{p_{k,1}} \mathbf{z} & \phi_k \mathbf{I} \end{bmatrix} \succeq \mathbf{0}, \exists \phi_k \geq 0. \tag{4.20}$$

Similarly, if there exists $\varphi_k \geq 0$, then the constraint (4.11d) can be reformulated as

$$\begin{bmatrix} \beta_k - \varphi_k & \mathbf{z}^H \hat{\mathbf{h}}_k \sqrt{p_{k,2}} & \mathbf{0} \\ \mathbf{z}^H \hat{\mathbf{h}}_k \sqrt{p_{k,2}} & 1 & \varepsilon_k \sqrt{p_{k,2}} \mathbf{z}^H \\ \mathbf{0} & \varepsilon_k \sqrt{p_{k,2}} \mathbf{z} & \varphi_k \mathbf{I} \end{bmatrix} \succeq \mathbf{0}, \exists \varphi_k \geq 0. \tag{4.21}$$

Therefore, the first subproblem (4.11) can be transformed as

$$\min_{\mathbf{z},\mathbf{U}_k,\alpha_k,\beta_k,\phi_k,\varphi_k,\forall k} \sum_{k=1}^{K} (\alpha_k + \beta_k) + \sigma_n^2 \|\mathbf{z}\|^2 \tag{4.22a}$$

$$\text{s.t.} \quad (5.11e), (4.11f), (4.15), (4.20), (4.21),$$

$$\phi_k \geq 0, \tag{4.22b}$$

$$\varphi_k \geq 0, \tag{4.22c}$$

$$\mathbf{U}_k \succeq \mathbf{0}, \tag{4.22d}$$

where the rank-one constraint $\text{Rank}(\mathbf{U}_k) = 1$ is omitted by applying the semi-definite relaxation (SDR) technique. It is seen that problem (4.22) is a joint convex problem, and thus it can be solved by some off-the-shelf optimization toolboxes, e.g., CVX [18]. Secondly, we address the other subproblem for transmit power allocation with the obtained solution $\{\mathbf{z}, \mathbf{U}_k\}$ from problem (4.22), which can be formulated as the following optimization problem:

$$\min_{p_{k,1},p_{k,2},\alpha_k,\beta_k,\forall k} \sum_{k=1}^{K} (\alpha_k + \beta_k) + \sigma_n^2 \|\mathbf{z}\|^2 \tag{4.23a}$$

$$\text{s.t.} \quad (5.17c), (5.17d), (5.17e), (4.11f), (4.15),$$

$$p_{k,1} + p_{k,2} \leq P_{\max,k}. \tag{4.23b}$$

To deal with the non-convexity of problem (4.23), we need to reformulate constraints (4.11c) and (4.11d). Now, we introduce a supplementary variable $P_{k,1} \triangleq \sqrt{p_{k,1}}$, then constraint (4.11c) can be rewritten as

$$\begin{bmatrix} \alpha_k - \phi_k & \mathbf{z}^H \hat{\mathbf{h}}_k P_{k,1} - 1 & \mathbf{0} \\ \mathbf{z}^H \hat{\mathbf{h}}_k P_{k,1} - 1 & 1 & \varepsilon_k P_{k,1} \mathbf{z}^H \\ \mathbf{0} & \varepsilon_k P_{k,1} \mathbf{z} & \phi_k \mathbf{I} \end{bmatrix} \succeq \mathbf{0}, \exists \phi_k \geq 0. \tag{4.24}$$

To reduce the design complexity, we utilize Lemma 4.1 to reformulate (4.11c) as

$$p_{k,2} \left(\hat{\mathbf{h}}_k^H \mathbf{z} \mathbf{z}^H \hat{\mathbf{h}}_k + 2\varepsilon_k \left\| \mathbf{z}\mathbf{z}^H \hat{\mathbf{h}}_k \right\| \right) \leq \beta_k. \tag{4.25}$$

Hence, the subproblem of transmit power allocation is transformed as

$$\min_{P_{k,1},p_{k,1},p_{k,2},\alpha_k,\beta_k,\phi_k,\forall k} \sum_{k=1}^{K} (\alpha_k + \beta_k) + \sigma_n^2 \|\mathbf{z}\|^2 \tag{4.26a}$$

$$\text{s.t.} \quad (5.17c), (5.17)d, (5.17r), (4.11f),$$

$$(4.15), (4.22b), (4.23b), (4.24), (4.25),$$

$$P_{k,1}^2 \leq p_{k,1}. \tag{4.26b}$$

It is obvious that problem (4.26) is a convex problem, which can be solved by CVX directly. In summary, the robust design of integrated computation and communication in B5G cellular IoT can be described as Algorithm 4.1.

Algorithm 4.1 Robust design of integrated computation and communication in B5G cellular IoT.

Input: N, K, σ_n^2, $P_{\max}$, ε_k, $\forall k$,
Output: $\mathbf{z}$, $\mathbf{u}_k$, $p_{k,1}$, $p_{k,2}$.
1: **Initialize** $\mathbf{z}$, $\mathbf{U}_k = \mathbf{u}_k \mathbf{u}_k^H$, $p_{k,1}^{(0)} = P_{\max,k}/2$, $p_{k,2}^{(0)} = P_{\max,k}/2$, and iteration index $t = 1$;
2: **repeat**
3:     obtain $\{\mathbf{z}^{(t)}, \mathbf{U}_k^{(t)}\}$ by solving problem (4.22) via CVX with fixed $\{p_{k,1}^{(t-1)}, p_{k,2}^{(t-1)}\}$;
4:     obtain $\{p_{k,1}^{(t)}, p_{k,2}^{(t)}\}$ by solving problem (4.26) via CVX with fixed $\{\mathbf{z}^{(t)}, \mathbf{U}_k^{(t)}\}$;
5:     $t = t + 1$;
6: **until** convergence
7: obtain $\mathbf{u}_k^{(t)}$ based on $\mathbf{U}_k^{(t)}$.

Remark 4.1 For the subproblem of optimizing receive beamforming (4.22), it is desired to recover the communication receiver $\mathbf{u}_k$ from the obtained $\mathbf{U}_k$. However, due to the rank relaxation, there is no guarantee that Rank$(\mathbf{U}_k) = 1$, $\forall k$. If $\mathbf{U}_k$ is a rank-one matrix, then $\mathbf{u}_k$ can be obtained by eigenvalue decomposition (EVD). Otherwise, the Gaussian randomization technique can be applied to obtain a suboptimal solution of communication receiver [22].

Complexity Analysis: The operation steps of each iteration in Algorithm 4.1 are the same, thus we only discuss its per-iteration complexity in the following. It is found that the main computational complexity comes from step 3, i.e., obtaining $\{\mathbf{z}, \mathbf{U}_k\}$ by optimizing problem (4.22) through a standard interior-point method (IMP) [19]. Specifically, it has $3K$ linear matrix inequality (LMI) constraints of size 1, $2K$ LMI constraints of size $N + 2$, and K LMI constraints of size N. Thus, for a given precision $\varepsilon > 0$ of solution, the per-iteration complexities of solving problem (4.22) by IMP is $\ln \frac{1}{\varepsilon} \sqrt{3K(N+3)} \cdot n \cdot \left[K\left(3 + 2(N+2)^3 + N^3\right) + Kn\left(3 + 2(N+2)^2 + N^2\right) \right]$, where the decision variable n is on the order of $\mathcal{O}(KM^2)$.

4.4 Numerical Results

In this section, we present simulation results to verify the robustness and effectiveness of the proposed Algorithm 4.1 in B5G cellular IoT. Without loss of generality, it is assumed that all IoT UEs are randomly distributed within a range of the cell radius, and have the same transmit power budget $P_{\max,k} = P_{\max}$. The path loss model is given by $\text{PL}_{dB} = 128.1 + 37.6 \log_{10}(d)$ [20], where d (km) is the distance between the BS and the IoT UE. For the sake of observation, we take normalized computation error

Table 4.2 Simulation parameters

Parameters	Values
BS antennas	$N = 32$
Number of IoT UEs	$K = 16$
Cell radius	500 (m)
Minimum required SINR threshold	$\gamma_k = \gamma_0 = 0.1$ dB
Noise power	$\sigma_n^2 = -70$ dBm
Region of channel uncertainty	$\varepsilon_k = \varepsilon = 0.1$

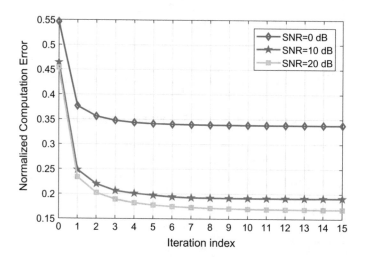

Fig. 4.2 Convergence behavior of the proposed Algorithm 4.1

MSE/K as the performance metric, and use SNR $= 10 \log_{10}(P_{\max}/\sigma_n^2)$ to denote the transmit SNR (in dB). Unless extra specification, the simulation parameters are set as in Table 4.2.

First, we show the convergence behaviors of the proposed Algorithm 4.1 under different SNR values in Fig. 4.2. It is shown that the normalized computation error gradually decreases as the number of iterations increases, and then stabilizes to an equilibrium point after no more than 10 iterations on average under different transmit SNR values. Combining with the complexity analysis for per-iteration, it is believed that the total complexity cost is affordable for practical implementation.

Then, Fig. 4.3 presents the performance comparison between the proposed Algorithm 4.1 and the non-robust algorithm. Note that the non-robust algorithm considers the obtained channel $\{\hat{\mathbf{h}}_k\}$ as the real channel in the design. It is seen that Algorithm 4.1 is always better than non-robust one in the whole region, and the gap between them enlarges with the increment of SNR, which validates the robustness and effectiveness of Algorithm 4.1.

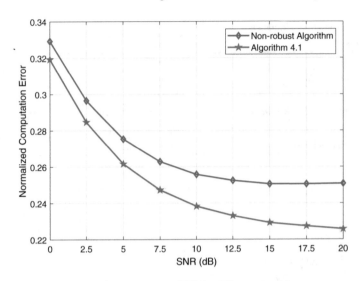

Fig. 4.3 Normalized computation error versus SNR for different algorithms

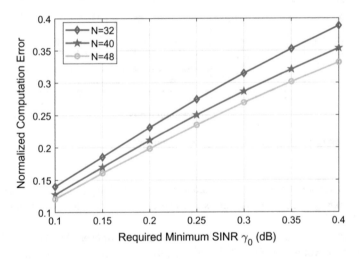

Fig. 4.4 Normalized computation error versus required minimum SINR for different number of BS antennas

Finally, we check the impacts of required minimum SINR at IoT UEs γ_0 and the number of BS antennas N on the computation error of Algorithm 4.1. As can be observed in Fig. 4.4, the normalized computation error grows as the required minimum SINR increases. This is because for a given transmit SNR at IoT UEs, the design with the higher required minimum SINR consumes more power to meet the communication performance, and incurs less power to decrease the computation error. Moreover, it is found that one can reduce the computation error by adding BS antennas, since more array gains can be exploited to improve the overall performance.

4.5 Conclusion

This chapter has designed a comprehensive framework for B5G cellular IoT integrating computation and communication. For realizing accurate computation and reliable communication with a massive of IoT UEs under practical conditions, a robust algorithm with the goal of minimizing the computation error was proposed by jointly optimizing transmit power and receive beamforming. Extensive simulations validated the effectiveness and robustness of the proposed algorithm for B5G cellular IoT integrating computation and communication.

A: The Proof of Lemma 4.1

According to the *Cauchy-Schwarz inequality* [21], we have

$$\left|\mathbf{p}^H\mathbf{q}\right| \leq \|\mathbf{p}\| \|\mathbf{q}\| \leq \delta \|\mathbf{q}\| . \tag{4.27}$$

Then, combining with the known inequality $-\left|\mathbf{p}^H\mathbf{q}\right| \leq \mathrm{Re}\{\mathbf{p}^H\mathbf{q}\} \leq \left|\mathbf{p}^H\mathbf{q}\right|$, we obtain

$$- \varepsilon \|\mathbf{q}\| \leq \mathrm{Re}\{\mathbf{p}^H\mathbf{q}\} \leq \delta \|\mathbf{q}\| . \tag{4.28}$$

To be specific, $f(\mathbf{p})$ achieves its upper bound $\delta \|\mathbf{q}\|$ and lower bound $-\delta \|\mathbf{q}\|$ at the points $\mathbf{p}_u = \frac{\delta}{\|\mathbf{q}\|}\mathbf{q}$ and $\mathbf{p}_l = -\frac{\delta}{\|\mathbf{q}\|}\mathbf{q}$, respectively. The proof is completed.

B: The Proof of Lemma 4.3

Based on the definition of semi-positive matrix, $\mathbf{F}(\mathbf{x}) \succeq \mathbf{0}, \forall \mathbf{x} : \|\mathbf{x}\| \leq \varepsilon$ holds true if and only if for $\forall \mathbf{d} \neq \mathbf{0}$,

$$\mathbf{d}^H\mathbf{A}\mathbf{d} - \mathbf{d}^H\left(\mathbf{B}^H\mathbf{x}\mathbf{c}^H + \mathbf{c}\mathbf{x}^H\mathbf{B}\right)\mathbf{d} \geq 0. \tag{4.29}$$

According to Lemma 4.2, (4.29) is equivalent to

$$\mathbf{d}^H\mathbf{A}\mathbf{d} \geq \max_{\|\mathbf{x}\|\leq\varepsilon} \mathbf{d}^H\left(\mathbf{B}^H\mathbf{x}\mathbf{c}^H + \mathbf{c}\mathbf{x}^H\mathbf{B}\right)\mathbf{d}$$
$$= 2\varepsilon|\mathbf{c}^H\mathbf{d}| \|\mathbf{B}\mathbf{d}\| . \tag{4.30}$$

Applying the *Cauchy-Schwarz inequality* [21], (4.30) can be further transformed as

$$\mathbf{d}^H \mathbf{A} \mathbf{d} - 2\varepsilon\varpi\mathbf{c}^H\mathbf{d} \geq 0, \forall \varpi : |\varpi| \leq \|\mathbf{B}\mathbf{d}\|. \tag{4.31}$$

Then, by exploiting *S-procedure* [17] with $\varpi^2 - \mathbf{d}^H \mathbf{B}^H \mathbf{B} \mathbf{d} \leq 0$, (4.31) is satisfied if and only if there exists a $\lambda \geq 0$ such that

$$\begin{bmatrix} \mathbf{A} - \lambda\mathbf{c}\mathbf{c}^H & -\varepsilon\mathbf{B}^H \\ -\varepsilon\mathbf{B} & \lambda\mathbf{I} \end{bmatrix} \succeq \mathbf{0}.$$

The proof is completed.

References

1. Chen X, Ng DWK, Yu W, Larsson EG, Al-Dhahir N, Schober R (2020) Massive access for 5G and beyond. IEEE J Sel Area Commun (99):1–24
2. Qi Q, Chen X, Zhong C, Zhang Z (2020) Robust integration of computation and communication in B5G cellular internet of things. In: Proceeding of IEEE WCNC, Seoul, South Korea, pp 1–6
3. Shirvanimoghaddam M, Dohler M, Johnson SJ (2017) Massive non-orthogonal multiple access for cellular IoT: potentials and limitations. IEEE Commun Mag 55(9):55–61
4. Chen X (2019) Massive access for cellular internet of things theory and technique. Springer, Germany
5. Qi Q, Chen X (2019) Wireless powered massive access for cellular internet of things with imperfect SIC and non-linear EH. IEEE Internet of Things J 6(2):3110–3120
6. Shirvanimogaddam M, Condoluci M, Dohler M, Johnson SJ (2017) On the fundamental limits of random non-orthogonal multiple access in cellular massive IoT. IEEE J Sel Areas Commun 35(10):2238–2252
7. Qi Q, Chen X, Lei L, Zhong C, Zhang Z (2019) Outage-constrained robust design for sustainable B5G cellular internet of things. IEEE Trans Wirel Commun 18(12):5780–5790
8. Chen X, Zhang Z, Zhong C, Jia R, Ng DWK (2018) Fully non-orthogonal communication for massive access. IEEE Trans Commun 16(4):6766–6778
9. Chen X, Zhang Z, Zhong C, Ng DWK (2017) Exploiting multiple-antenna techniques for non-orthogonal multiple access. IEEE J Sel Areas Commun 35(10):2207–2220
10. Zihan E, Choi KW, Kim DI (2015) Distributed random access scheme for collision avoidance in cellular device-to-device communication. IEEE Trans Wirel Commun 14(7):3571–3585
11. Abari O, Rahul H, Katabi D (2016) Over-the-air function computation in sensor networks. http://arxiv.org/pdf/1612.02307.pdf
12. Goldenbaum M, Boche H, Staczak S (2013) Harnessing interference for analog function computation in wireless sensor networks. IEEE Trans Sig Process 61(204):4893–4906
13. Zhu G, Huang K (2019) MIMO over-the-air computation for high-mobility multimodal sensing. IEEE Internet of Things J 6(4):6089–6103
14. Qi Q, Chen X, Lei L, Zhong C, Zhang Z (2019), Robust convergence of energy and computation for B5G cellular internet of things. In: Proceeding of IEEE GLOBECOM, Hawaii, USA, pp 1–6
15. Goldenbaum M, Boche H, Staczak S (2015) Nomographic functions: efficient computation in clustered gaussian sensor networks. IEEE Trans Wirel Commun 14(4):2093–2105
16. Wang J, Palomar DP (2009) Worst-case robust MIMO transmission with imperfect channel knowledge. IEEE Trans Sig Process 57(8):3086–3100

17. Boyd S, Vandenberghe L (2004) Convex Optimization. Cambridge University Press, Cambridge, UK
18. Grant M, Boyd S. CVX: Matlab software for disciplined convex programming. http://cvxr.com/cvx
19. Ben-Tal A, Nemirovski A (2001) Lectures on modern convex optimization: analysis, algorithms, and engineering applications. In: SIAM, MPS-SIAM series on optimization, Philadelphia, PA, USA
20. 3GPP, Coordinated multi-point operation for LTE physical layer aspects (Release 11) (2011)
21. Hardy GH, Littlewood JE, Polya G (1952) Inequalities. Cambridge University Press, Cambridge, UK
22. Luo Z, Ma W, So AM, Ye Y, Zhang S (2010) Semidefinite relaxation of quadratic optimization problems. IEEE Sig Process Mag 27(3):20–34

Chapter 5
Convergence of Energy, Communication and Computation in B5G Cellular Internet of Things

Abstract In this chapter, in order to jointly address the critical issues of B5G cellular IoT, i.e., energy supply, data aggregation, and information transmission, we design a framework integrating energy, computation and communication (ECC). Firstly, the BS charges massive IoT devices simultaneously via the WPT technique in the downlink. Then, IoT devices with harvested energy carry out the computation task and the communication task in the uplink via AirComp and non-orthogonal transmission over the same spectrum. To improve the overall performance of ECC, we propose a joint beamforming design algorithm for the BS and the IoT devices. Finally, simulation results validate the effectiveness of the proposed algorithm in B5G cellular IoT.

5.1 Introduction

In general, IoT devices equipped with sensors collect information from the surroundings or human, and then transmit it to the BS for further decoding, computation, or analysis [1]. The signals from the IoT devices usually have two functions, one for data aggregation based on multiple devices' signals, and the other for information transmission based on individual device's signal. Thus, computation and communication can be abstracted as two elementary tasks of B5G cellular IoT. However, it is not trivial to carry out the two tasks with limited wireless resources. Specifically, for computation, the conventional way of *transmit-then-compute* is not suitable for massive IoT and data, due to the ultra-high latency and the low spectrum utilization. To tackle this issue, a promising solution called AirComp has been proposed in [2, 3], which takes advantage of the superposition property of the wireless multiple-access channels (MAC) to compute a class of *nomographic functions* [4] of distributed data from IoT devices by concurrent transmission. Especially in B5G cellular IoT, Air-Comp can combine with multiple-input and multiple-output (MIMO) techniques to spatially multiplex multi-function computation, and further decreases the computation errors by using spatial beamforming [5, 6].

© The Author(s), under exclusive license to Springer Nature Singapore Pte Ltd. 2020   111
X. Chen and Q. Qi, *Convergence of Energy, Communication and Computation in B5G Cellular Internet of Things*, SpringerBriefs in Electrical and Computer Engineering, https://doi.org/10.1007/978-981-15-4140-7_5

On the other hand, for communication, traditional orthogonal multiple access (OMA) schemes cannot support massive access due to limited radio spectrum. In this case, non-orthogonal multiple access (NOMA) is applied to B5G cellular IoT to realize seamless access of a massive number of devices [7, 8]. However, massive NOMA incurs severe co-channel interference degrading the quality of communication signals [9]. In this case, spatial beamforming is usually exploited to combat co-channel interference as well as improve the system performance [10, 11]. In particular, since the BS of B5G cellular IoT is equipped with a large-scale antenna array, there are ultra-high spatial degrees of freedom to mitigate co-channel interference [12, 13].

Besides, energy supply for a massive number of IoT devices is a crucial problem in B5G cellular IoT. Due to the high human cost and the environmental strain, frequent battery replacement for massive IoT is prohibitive. Hence, it is appealing to adopt the wireless power transfer (WPT) technique to realize one-to-many charging by exploiting the open nature of the wireless broadcast channels [14]. To improve the efficiency of WPT over fading channels, energy beamforming has been introduced in [15–17].

In fact, energy supply, data aggregation, and information transmission are not separated. They compete the same wireless resources for performance enhancement. To improve the overall performance with limited wireless resources, this chapter designs a framework integrating energy, computation, and communication (ECC) for B5G cellular IoT. The contributions of this chapter are two-fold:

1. We put forward a sustainable design framework for the integration of ECC in B5G cellular IoT, including energy supply by WPT in the downlink, and computation by AirComp and communication by NOMA in the uplink.
2. We analyze the impacts of transmit and receive beamforming on the overall performance, and propose an algorithm to minimize the computation distortion while ensuring the communication requirements via power allocation and beamforming design.

The rest of this chapter is organized as follows: Sect. 5.2 gives a concise introduction of B5G cellular IoT integrating ECC. Section 5.3 focuses on the algorithm design to improve the overall performance. Section 5.4 presents simulation results to validate the effectiveness of the proposed algorithm. Finally, Sect. 5.5 concludes the chapter.

Notations: We use bold upper (lower) letters to denote matrices (column vectors), $(\cdot)^H$ to denote conjugate transpose, $\|\cdot\|_F$ to denote Frobenius norm of a matrix, $\|\cdot\|$ to denote L_2-norm of a vector, $\mathbb{E}\{\cdot\}$ to denote expectation, $\mathrm{tr}(\cdot)$ to denote trace of a matrix, $\mathrm{Rank}(\cdot)$ to denote rank of a matrix, $\mathbb{C}^{m \times n}$ to denote the set of m-by-n dimensional complex matrix.

5.2 System Model

Let us consider a sustainable B5G cellular IoT network as shown in Fig. 5.1, where a BS at the cell center equipped with N antennas communicates with K multi-modal IoT user equipments (UEs) equipped with M antennas each. The system is operated in TDD mode. During the first half of a time slot, the BS acts as a power beacon to charge IoT UEs by energy beamforming. The harvested energy at the kth UE can be expressed as

$$E_k = \frac{T}{2}\vartheta_k \left\| \mathbf{H}_k^H \mathbf{f} \right\|^2, \tag{5.1}$$

where $\mathbf{H}_k \in \mathbb{C}^{N \times M}$ denotes the MIMO channel matrix from the kth UE to the BS, which remains constant in a time slot, but independently fades over time slots. T denotes the length of a time slot, ϑ_k denotes the energy conversion efficiency, and $\mathbf{f} \in \mathbb{C}^{N \times 1}$ denotes the energy beam transmitted by the BS.

After harvesting enough energy in the downlink, IoT UEs are required to complete two fundamental tasks in the uplink, namely computation and communication. Specifically, in the second half of the time slot, each IoT UE performs beamforming on the computation signal and the communication signal to be transmitted respectively, and then sends a superposition coded signal to the BS over the uplink channel.

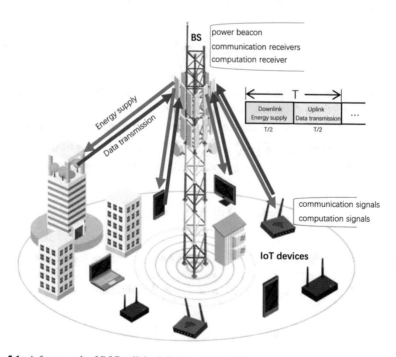

Fig. 5.1 A framework of B5G cellular IoT integrating ECC

Table 5.1 Some examples of nomographic functions

Functions	Pre-processing	Post-processing	Targeted function
Arithmetic mean	$g_k = d_k$	$f = 1/K$	$q = \frac{1}{K}\sum_{k=1}^{K} d_k$
Weighted sum	$g_k = \eta_k d_k$	$f = 1$	$q = \sum_{k=1}^{K} \eta_k d_k$
Geometric mean	$g_k = \ln(d_k)$	$f = \exp(\cdot)$	$q = \left(\prod_{k=1}^{K} d_k\right)^{1/K}$
Polynomial	$g_k = \eta_k d_k^{\beta_k}$	$f = 1$	$q = \sum_{k=1}^{K} \vartheta_k d_k^{\beta_k}$
Euclidean norm	$g_k = d_k^2$	$f = \sqrt{(\cdot)}$	$q = \sqrt{\sum_{k=1}^{K} d_k^2}$

On the one hand, enabled by AirComp, the BS obtains the computation results directly via concurrent data transmission, and then designs a computation receiver to recover the targeted function signal. On the other hand, the BS decodes the communication signals of each UE through communication receivers.

Without loss of generality, we assume that the kth UE records data of L heterogeneous parameters to be computed and J heterogeneous parameters to be communicated, which generate a computation symbol vector $\mathbf{d}_k = \left[d_{k,1}, \ldots, d_{k,L}\right]^T$ and a communication symbol vector $\mathbf{s}'_k = \left[s'_{k,1}, \ldots, s'_{k,J}\right]^T$, where $d_{k,l}$ and $s'_{k,j}$ are the measured values of the lth computation parameter and the jth communication parameter at the kth UE, respectively. Note that for computation, the BS engages itself in computing L nomographic functions via AirComp [2, 3], such that

$$q_l = f_l \left(\sum_{k=1}^{K} g_{k,l}\left(d_{k,l}\right)\right), l = 1, ..., L, \tag{5.2}$$

where $g_{k,l}(\cdot)$ and $f_l(\cdot)$ represent pre-processing functions at the IoT UEs and post-processing functions at the BS, respectively (see Table 5.1 for examples). Let $\mathbf{s}_k = \left[g_{k,1}\left(d_{k,1}\right), \ldots, g_{k,L}\left(d_{k,L}\right)\right]^T$ denote the pre-processed computation signal at the kth UE. Here, we assume that $\mathbb{E}\left\{\mathbf{s}_k \mathbf{s}_k^H\right\} = \mathbf{I}$ and $\mathbb{E}\left\{s'_{k,j} s'^H_{k,j}\right\} = 1$ for ease of analysis. Thus, the kth UE constructs the superposition coded transmit signal $\mathbf{x}_k$ as follows

$$\mathbf{x}_k = \mathbf{W}_k \mathbf{s}_k + \sum_{j=1}^{J} \mathbf{v}_{k,j} s'_{k,j}, \tag{5.3}$$

where $\mathbf{W}_k \in \mathbb{C}^{M \times L}$ and $\mathbf{v}_{k,j} \in \mathbb{C}^{M \times 1}$ denote the transmit beams for the computation signal and the communication signal, respectively. Note that the transmit power of each IoT UE comes from the harvested energy in the downlink, namely

$$\|\mathbf{W}_k\|_F^2 + \sum_{j=1}^{J} \|\mathbf{v}_{k,j}\|^2 \leq \frac{E_k}{T/2}, \forall k, j. \tag{5.4}$$

Therefore, the received signal at the BS is given by

$$\mathbf{y} = \sum_{k=1}^{K} \mathbf{H}_k \mathbf{x}_k + \mathbf{n}$$

$$= \underbrace{\sum_{k=1}^{K} \mathbf{H}_k \mathbf{W}_k \mathbf{s}_k}_{\text{computation signal}} + \underbrace{\sum_{k=1}^{K} \sum_{j=1}^{J} \mathbf{H}_k \mathbf{v}_{k,j} s'_{k,j}}_{\text{communication signal}} + \mathbf{n}, \tag{5.5}$$

where $\mathbf{n}$ is the additive white Gaussian noise (AWGN) vector with the variance σ_n^2. Firstly, we discuss the processing of the computation signals. Due to the one-to-one mapping between $\mathbf{s} = \sum_{k=1}^{K} \mathbf{s}_k$ and $\mathbf{q} = [q_1, q_2, ..., q_L]^T$ in (5.2), we consider an accurate $\mathbf{s}$ at the BS as the targeted function signal. To minimize the distortion of the targeted function signal caused by channel fading, interference, and noise, it is desired to perform receive beamforming at the BS. Thus, the received signal for computation at the BS is given by

$$\hat{\mathbf{s}} = \mathbf{Z}^H \sum_{k=1}^{K} \mathbf{H}_k \mathbf{W}_k \mathbf{s}_k + \mathbf{Z}^H \left(\sum_{k=1}^{K} \mathbf{H}_k \sum_{j=1}^{J} \mathbf{v}_{k,j} s'_{k,j} + \mathbf{n} \right), \tag{5.6}$$

where $\mathbf{Z} \in \mathbb{C}^{N \times L}$ is a receive beam for computation results at the BS. As a rule, the performance of AirComp at the BS is measured by the mean square error (MSE) between $\mathbf{s}$ and $\hat{\mathbf{s}}$, which is defined as

$$\text{MSE} \left(\hat{\mathbf{s}}, \mathbf{s} \right) = \mathbb{E} \left\{ \text{tr} \left(\left(\hat{\mathbf{s}} - \mathbf{s} \right) \left(\hat{\mathbf{s}} - \mathbf{s} \right)^H \right) \right\}. \tag{5.7}$$

Substituting (5.6) into (5.7), the computation distortion can be expressed as the following MSE function of receive and transmit beams:

$$\text{MSE} \left(\mathbf{Z}, \mathbf{W}_k, \mathbf{v}_{k,j} \right) = \sum_{k=1}^{K} \left\| \mathbf{Z}^H \mathbf{H}_k \mathbf{W}_k - \mathbf{I} \right\|_F^2 + \sigma_n^2 \left\| \mathbf{Z} \right\|_F^2$$

$$+ \sum_{k=1}^{K} \sum_{j=1}^{J} \left\| \mathbf{Z}^H \mathbf{H}_k \mathbf{v}_{k,j} \right\|^2. \tag{5.8}$$

Secondly, we handle the processing of the communication signals. The received signal for communication at the BS can be expressed as

$$y'_{k,j} = \mathbf{u}^H_{k,j}\mathbf{H}_k\mathbf{v}_{k,j}s'_{k,j} + \mathbf{u}^H_{k,j}\sum_{i=1,i\neq k}^{K}\mathbf{H}_i\sum_{m=1,m\neq j}^{J}\mathbf{v}_{i,m}s'_{i,m}$$

$$+ \mathbf{u}^H_{k,j}\sum_{i=1}^{K}\mathbf{H}_i\mathbf{W}_i\mathbf{s}_i + \mathbf{u}^H_{k,j}\mathbf{n}, \tag{5.9}$$

where $\mathbf{u}_{k,j} \in \mathbb{C}^{N\times 1}$ denotes the receive beamforming vector for the communication signal $s'_{k,j}$ at the BS. As a consequence, the received signal-to-interference-plus-noise ratio (SINR) at the communication receiver can be expressed as

$$\Gamma_{k,j} = \frac{\left|\mathbf{u}^H_{k,j}\mathbf{H}_k\mathbf{v}_{k,j}\right|^2}{\displaystyle\sum_{i=1,i\neq k}^{K}\sum_{m=1,m\neq j}^{J}\left|\mathbf{u}^H_{k,j}\mathbf{H}_i\mathbf{v}_{i,m}\right|^2 + \sum_{i=1}^{K}\left\|\mathbf{u}^H_{k,j}\mathbf{H}_i\mathbf{W}_i\right\|^2 + \sigma_n^2\left\|\mathbf{u}_{k,j}\right\|^2}. \tag{5.10}$$

As seen from (5.8) and (5.10), the system performance depends on the transmit beams $\mathbf{W}_k$ and $\mathbf{v}_{k,j}$ at the IoT UEs, and receive beams $\mathbf{Z}$ and $\mathbf{u}_{k,j}$ at the BS. Moreover, the transmit power of IoT UEs depnds on the energy beam sent by the BS. Thus, it makes sense to jointly design transmit and receive beamforming of ECC for enhancing the overall performance of B5G cellular IoT.

5.3 Problem Formulation and Optimization Solution

In this section, we design an algorithm to realize an efficient integration of ECC in B5G cellular IoT. The design aims to minimize the computation distortion, while guaranteeing the SINR requirements of communication signals, which can be formulated as the following optimization problem:

$$\min_{\substack{\mathbf{W}_k, \mathbf{v}_{k,j}, \\ \mathbf{f}, \mathbf{u}_{k,j}, \mathbf{Z}}} \quad \text{MSE}\left(\mathbf{W}_k, \mathbf{v}_{k,j}, \mathbf{Z}\right) \tag{5.11a}$$

$$\text{s.t.} \quad \Gamma_{k,j} \geq \gamma_{k,j}, \tag{5.11b}$$

$$\|\mathbf{W}_k\|_F^2 + \sum_{j=1}^{J}\|\mathbf{v}_{k,j}\|^2 \leq \vartheta_k\|\mathbf{H}_k^H\mathbf{f}\|^2, \tag{5.11c}$$

$$\|\mathbf{f}\|^2 \leq P_{\max}, \tag{5.11d}$$

where $\gamma_{k,j}$ is the required minimum SINR of the jth communication signal at the kth UE, and $P_{\max}$ is the maximum transmit power budget at the BS for the energy beam. Problem (5.11) is non-convex due to the coupled variables. To tackle this issue, we adopt an alternating optimization (AO) algorithm by dividing the original problem into two subproblems, one for the optimization of receive beams, and the other for the optimization of transmit beams. The AO algorithm stops until the objective value of the original problem approaches a stationary point in the iterations. Now, we first consider the subproblem for the optimization of receive beams, i.e., $\{\mathbf{Z}, \mathbf{u}_{k,j}\}$. To balance computational complexity and system performance, we employ the minimum mean square error (MMSE) receivers, which are given by

$$\mathbf{Z} = \left(\sigma_n^2 \mathbf{I} + \sum_{k=1}^{K} \mathbf{H}_k \Xi_k \mathbf{H}_k^H\right)^{-1} \sum_{k=1}^{K} \mathbf{H}_k \mathbf{W}_k, \tag{5.12}$$

and

$$\mathbf{u}_{k,j} = \left(\sigma_n^2 \mathbf{I} + \sum_{k=1}^{K} \mathbf{H}_k \Xi_k \mathbf{H}_k^H\right)^{-1} \mathbf{H}_k \mathbf{v}_{k,j}, \tag{5.13}$$

respectively, where $\Xi_k = \mathbf{W}_k \mathbf{W}_k^H + \sum_{j=1}^{J} \mathbf{v}_{k,j} \mathbf{v}_{k,j}^H$. Next, we deal with the other subproblem for the optimization of transmit beams $\{\mathbf{W}_k \ \mathbf{v}_{k,j}, \mathbf{f}\}$ with fixed MMSE receivers $\{\mathbf{Z}, \mathbf{u}_{k,j}\}$ in (5.12) and (5.13). To address the non-convexity of constraints (5.11b) and (5.11c), we introduce $\mathbf{V}_{k,j} = \mathbf{v}_{k,j} \mathbf{v}_{k,j}^H$ and $\mathbf{F} = \mathbf{f}\mathbf{f}^H$. Thus, the SINR constraint (5.11b) can be transformed as

$$\frac{1}{\gamma_{k,j}} \left(\mathbf{u}_{k,j}^H \mathbf{H}_k \mathbf{V}_{k,j} \mathbf{H}_k^H \mathbf{u}_{k,j}\right) \geq \sum_{i=1}^{K} \left\|\mathbf{u}_{k,j}^H \mathbf{H}_i \mathbf{W}_i\right\|^2$$

$$+ \sum_{i=1, i\neq k}^{K} \sum_{m=1, m\neq j}^{J} \left(\mathbf{u}_{k,j}^H \mathbf{H}_i \mathbf{V}_{i,m} \mathbf{H}_i^H \mathbf{u}_{k,j}\right) + \sigma_n^2 \left\|\mathbf{u}_{k,j}\right\|^2. \tag{5.14}$$

Accordingly, the power constraints of transmit beams (5.11c) and (5.11d) can be rewritten as

$$\|\mathbf{W}_k\|_F^2 + \sum_{j=1}^{J} \mathrm{tr}\left(\mathbf{V}_{k,j}\right) \leq \vartheta_k \mathrm{tr}\left(\mathbf{H}_k^H \mathbf{F} \mathbf{H}_k\right), \tag{5.15}$$

and

$$\mathrm{tr}\left(\mathbf{F}\right) \leq P_{\max}, \tag{5.16}$$

respectively. Then, the subproblem can be formulated as the following semi-definite programming (SDP) problem:

$$\min_{\mathbf{W}_k, \mathbf{V}_{k,j}, \mathbf{F}} \quad \sum_{k=1}^{K} \left\| \mathbf{Z}^H \mathbf{H}_k \mathbf{W}_k - \mathbf{I} \right\|_F^2$$

$$+ \sum_{k=1}^{K} \sum_{j=1}^{J} \mathrm{tr}\left(\mathbf{Z}^H \mathbf{H}_k \mathbf{V}_{k,j} \mathbf{H}_k^H \mathbf{Z} \right) \tag{5.17a}$$

$$\text{s.t.} \quad (5.14), (5.15), (5.16),$$

$$\mathbf{V}_{k,j} \succeq \mathbf{0}, \forall k, j, \tag{5.17b}$$

$$\mathbf{F} \succeq \mathbf{0}, \tag{5.17c}$$

$$\mathrm{rank}(\mathbf{V}_{k,j}) = 1, \forall k, j, \tag{5.17d}$$

$$\mathrm{rank}(\mathbf{F}) = 1. \tag{5.17e}$$

Since problem (5.17) dropping rank-one constraints (5.17d) and (5.17e) is a convex problem, it can be effectively solved by some optimization tools, such as CVX [18]. For the obtained solutions $\{\mathbf{V}_{k,j}^*, \mathbf{F}^*\}$ to problem (5.17), we have the following theorem.

Theorem 5.1 *The optimal solutions* $\{\mathbf{V}_{k,j}^*, \mathbf{F}^*\}$ *of problem (5.17) always satisfy rank-one constraints* $\mathrm{Rank}\left(\mathbf{V}_{k,j}^*\right) = 1, \forall k, j$ *and* $\mathrm{rank}(\mathbf{F}^*) = 1$.

Proof Due to the similar analysis in [11], we only give a brief proof thought here. First, we construct the lagrangian function of problem (5.17). Then, we reveal the structure of the optimal solutions $\{\mathbf{V}_{k,j}^*, \mathbf{F}^*\}$ by exploiting the Slaters condition and the Karush-Kuhn-Tucher (KKT) conditions. Finally, according to the listed KKT conditions, we deduce the rank-one relationship of obtained solutions by some rank inequalities. The detailed derivation process can be referred to [11].

Hence, we can recover the unique transmit communication beams $\mathbf{v}_{k,j}^*$ and energy beam $\mathbf{f}^*$ of the original problem (5.11) via eigenvalue decomposition (EVD). By alternately optimizing the two subproblems, we develop an iterative algorithm which always converges to a stationary point as the objective value decreases in the iterations. In summary, the design of B5G cellular IoT integrating ECC for minimizing the computation error can be described as Algorithm 5.1.

Complexity Analysis: It is seen from Algorithm 5.1 that the main computational complexity comes from step 5, i.e., solving problem (5.17). Since problem (5.17) only contains linear matrix inequality (LMI) and second-order cone (SOC) constraints, it can be solved by a standard interior-point method (IPM) [19]. Specifically, there are 2 LMI constraints of size N, KJ LMI constraints of size M, and $K(J+1)$ SOC constraints of size M. Thus, for a given precision $\varepsilon > 0$, the per-iteration complexity of solving problem (5.17) by IPM is $\sqrt{(2+M)KJ + 2(N+K)} \cdot \ln(1/\varepsilon) \cdot n \cdot \left[2N^3 + KJM^3 + n\left(2N^2 + KJM^2\right) + (J+1)KM^2 + n^2\right]$, where the decision variable n is on the order of $\mathcal{O}(KM^2)$.

Algorithm 5.1 Design of B5G Cellular IoT integrating ECC for the Computation Error Minimization.

Input: $N, K, M, L, J, \sigma_n^2, \gamma_{k,j}, P_{\max}, \forall k, j$
Output: $\mathbf{W}_k, \mathbf{v}_{k,j}, \mathbf{Z}, \mathbf{u}_{k,j}, \mathbf{f}, \forall k, j$
1: **Initialize** $\mathbf{W}_k^{(0)}, \mathbf{v}_{k,j}^{(0)} \forall k, j$, iteration index $t = 1$;
2: **repeat**
3: compute $\mathbf{Z}^{(t)}$ by (5.12) with $\mathbf{W}_k^{(t-1)}$ and $\mathbf{v}_{k,j}^{(t-1)}$;
4: compute $\mathbf{u}_{k,j}^{(t)}$ by (5.13) with $\mathbf{W}_k^{(t-1)}$ and $\mathbf{v}_{k,j}^{(t-1)}$;
5: obtain $\{\mathbf{W}_k^{(t)}, \mathbf{V}_{k,j}^{(t)}, \mathbf{F}^{(t)}\}$ by solving problem (5.17) via CVX with $\{\mathbf{Z}^{(t)}, \mathbf{u}_{k,j}^{(t)}\}$;
6: obtain $\mathbf{v}_{k,j}^{(t)}$ and $\mathbf{f}^{(t)}$ via EVD;
7: update $t = t + 1$;
8: **until** convergence

5.4 Simulation Results

In this section, we provide several simulation results to validate the effectiveness of the proposed algorithm. Without loss of generality, it is assumed that all IoT UEs are randomly distributed in a cell with a radius R. The pass loss is modeled as $PL_{dB} = 128.1 + 37.6 \log_{10}(d)$ [20], where d (km) is the distance between the BS and the IoT UE. For ease of analysis, we take the normalized computation error MSE/K as the AirComp performance metric, and use $SNR = 10 \log_{10}(P_{\max}/\sigma_n^2)$ to denote the transmit signal-to-noise ratio (SNR) (in dB). Unless otherwise stated, the simulation parameters are listed in Table 5.2.

First, we present the convergence behaviors of the proposed Algorithm 5.1 with different SNRs in Fig. 5.2. It is found that Algorithm 5.1 has a quick convergence at higher SNR, but needs more iterations at lower SNR. This is because there exists severe co-channel interference in the received signals at the BS for a small SNR, which affects the performance. More importantly, Algorithm 5.1 always converges with no more than 10 iterations under different SNRs, which implies that the computational complexity is tolerable for practical implementation.

Next, we show the performance comparison among Algorithm 5.1 and three baseline beamforming design schemes in Fig. 5.3, i.e., a fixed MMSE scheme whose

Table 5.2 Simulation parameters

Parameters	Values
Number of BS antennas	$N = 64$
IoT UEs	$K = 32, M = 2, L = 1, J = 1$
Energy conversion efficiency	$\vartheta_k = \vartheta_0 = 0.5$
Cell radius	$R = 500$ m
Minimum required SINR threshold	$\gamma_{k,j} = \gamma_0 = 0.1$ dB
Noise power	$\sigma_n^2 = -50$ dBm

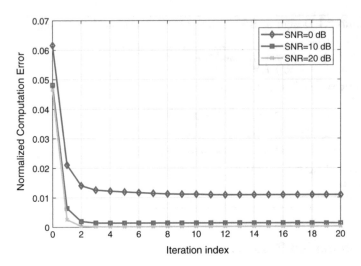

Fig. 5.2   Convergence behavior of the proposed Algorithm 5.1

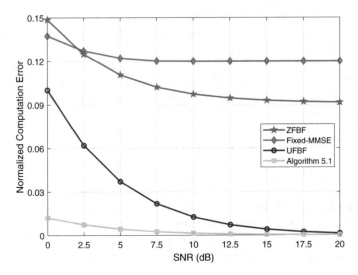

Fig. 5.3   Normalized computation error versus SNR (dB) for three beamforming design algorithms

receivers are only related to the channels, a zero-forcing beamforming (ZFBF) scheme whose transmitters are designed based on the zero-forcing principle, and an uniform-forcing beamforming (UFBF) scheme based on the AO algorithm with uniform-forcing transmitters and MMSE receivers. As the SNR increases, the computation error decreases for these four schemes. It is seen that the ZFBF scheme performs worse than the fixed MMSE scheme in the low SNR region, but has a performance advantage in the high SNR region. UFBF scheme is far ahead of these

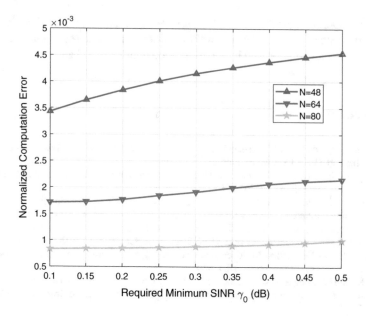

Fig. 5.4 Normalized computation error versus required minimum SINR (dB) for different numbers of BS antennas with the proposed Algorithm 5.1

two algorithms in the whole SNR region, but is always behind the proposed Algorithm 5.1, especially for the low and medium regions. This is because Algorithm 5.1 jointly optimizes the receive and transmit beamforming adaptively to channel conditions and system parameters.

Finally, Fig. 5.4 illustrates the influences of the required minimum SINR γ_0 and the number of BS antennas N on the performance of proposed Algorithm 5.1 with SNR = 10 dB. It is seen that the normalized computation error increases with the increment of the required minimum SINR, since more power is used to enhance the quality of the communication signals, resulting in less power consumed to reduce the computation distortion. Besides, the computation error decreases as the number of BS antennas increases. This is because more array gains can be obtained to improve the overall performance.

5.5 Conclusion

This chapter has designed a sustainable framework for B5G cellular IoT integrating ECC. For realizing accurate computation and efficient communication with the harvested energy at a massive number of IoT UEs, a joint beamforming design algorithm was proposed from the perspective of minimizing the computation error while ensuring the SINR requirements of communication signals. It was revealed that the proposed algorithm was able to effectively integrate ECC and achieved the best performance over baseline algorithms. Moreover, it was found that the overall performance can be improved by increasing the number of BS antennas.

References

1. Chen X, Ng DWK, Yu W, Larsson EG, Al-Dhahir N, Schober R (2020) Massive access for 5G and beyond. IEEE J Sel Area Commun (99):1–24
2. Abari O, Rahul H, Katabi D (2016) Over-the-air function computation in sensor networks. http://arxiv.org/pdf/1612.02307.pdf
3. Chen L, Zhao N, Chen Y, Yu FR, Wei G (2018) Over-the-air computation for IoT networks: computing multiple functions with antenna arrays. IEEE Internet of Things J 5(6):5296–5306
4. Goldenbaum M, Boche H, Staczak S (2015) Nomographic functions: efficient computation in clustered gaussian sensor networks. IEEE Trans Wirel Commun 14(4):2093–2105
5. Zhu G, Huang K (2019) MIMO over-the-air computation for high-mobility mult-modal sensing. IEEE Internet of Things J 6(4):6089–6103
6. Li X, Zhu G, Gong Y, Huang K (2019) Wirelessly powered data aggregation for IoT via over-the-air function computation: beamforming and power vontrol. IEEE Trans Wirel Commun 18(7):3437–3452
7. Chen X (2019) Massive access for cellular internet of things theory and technique. Springer, Germany
8. Chen X, Zhang Z, Zhong C, Jia R, Ng DWK (2018) Fully non-orthogonal communication for massive access. IEEE Trans Commun 16(4):6766–6778
9. Shirvanimoghaddam M, Dohler M, Johnson SJ (2017) Massive non-orthogonal multiple access for cellular IoT: potentials and limitations. IEEE Commun Mag 55(9):55–61
10. Jia R, Chen X, Zhong C, Ng DWK, Lin H, Zhang Z (2019) Design of non-orthogonal beamspace multiple access for cellular internet-of-things. IEEE J Sel Top Sig Process 13(3):538–552
11. Qi Q, Chen X (2019) Wireless powered massive access for cellular internet of things with imperfect SIC and non-linear EH. IEEE Internet of Things J 6(2):3110–3120
12. Chen X, Zhang Z, Zhong C, Ng DWK (2017) Exploiting multiple-antenna techniques for non-orthogonal multiple access. IEEE J Sel Areas Commun 35(10):2207–2220
13. Chen X, Jia R, Ng DWK (2019) On the design of massive non-orthogonal multiple access with imperfect successive interference cancellation. IEEE Trans Commun 67(3):2539–2551
14. Chen X, Zhang Z, Chen H-H, Zhang H (2015) Enhancing wireless information and power transfer by exploiting multi-antenna techniques. IEEE Commun Mag 53(4):133–141
15. Qi Q, Chen X, Lei L, Zhong C, Zhang Z (2019) Outage-constrained robust design for sustainable B5G cellular internet of things. IEEE Trans Wirel Commun 18(12):5780–5790
16. Chen X, Yuen C, Zhang Z (2014) Wireless energy and information transfer tradeoff for limited feedback multi-antenna systems with energy beamforming. IEEE Trans Vehic Technol 63(1):407–412
17. Chen X, Wang X, Chen X (2013) Energy-efficient optimization for wireless information and power transfer in large-scale MIMO systems employing energy beamforming. IEEE Wirel Commun Lett 2(6):667–670
18. Grant M, Boyd S. CVX: Matlab software for disciplined convex programming. http://cvxr.com/cvx
19. Ben-Tal A, Nemirovski A (2001) Lectures on modern convex optimization: analysis, algorithms, and engineering applications. In: SIAM, MPS-SIAM Series on Optimization, Philadelphia, PA, USA
20. 3GPP, Coordinated multi-point operation for LTE physical layer aspects (Release 11) (2011)

Chapter 6
Summary

Abstract In this chapter, we make a summary about the convergence of energy, communication and computation in B5G cellular IoT. At first, we comprehensively and systematically discuss the key techniques of energy, communication and computation in B5G cellular IoT, including wireless power transfer, non-orthogonal multiple access, over-air-computation, and massive multiple-input multiple-output. Four typical convergence scenarios are studied in detail, namely convergence of energy and communication, convergence of energy and computation, convergence of communication and computation, and convergence of energy, communication and computation. In particular, we provide in-depth design, analysis and optimization for each convergence scenario of energy, communication and computation in B5G cellular IoT. In addition, we analyze the challenging issues in the existing schemes about energy, communication and computation in B5G cellular IoT, and point out the future research directions for further improving the overall performance of B5G cellular IoT.

6.1 Concluding Remarks

Nowadays, the rapid development of IoT incurs the explosive growth in the number of terminal devices and the surge of data traffic. In order to support real-time processing of mass data of IoT devices with multiple tasks, B5G cellular IoT has to be a large-scale edge-intelligent network to meet the requirements of ultra-low latency, ultra-high efficiency, ultra-high reliability and ultra-high density connectivity. In specific, energy, communication and computation are carried out at the edge of cellular IoT by exploiting the potential of massive IoT devices. Hence, traditional frameworks and techniques for cellular IoT which process data at cloud servers is not applicable any more. In this context, several promising techniques, such as wireless power transfer, non-orthogonal multiple access, over-air-computation, and massive multiple-input multiple-output, are applied into B5G cellular IoT. In this book, we propose four general frameworks which integrates energy, communication and computation at the edge of cellular networks by making use of the open nature of wireless channels.

To enhance the overall system performance, corresponding schemes including design, analysis and optimization are provided. The main contributions of this book are summarized as follows.

In Chap. 1, we first introduce the origin and development of cellular IoT. Then, we discuss the characteristics of B5G cellular IoT. In general, IoT devices equipped with advanced sensors collect information from the surroundings or human, then transmit it to the BS for further decoding, computation, or analysis. The signals from the IoT devices usually have two functions, one for data aggregation based on multiple devices' signals, and the other for information transmission based on individual device's signal. Thus, computation and communication can be abstracted as two elementary tasks of B5G cellular IoT. However, it is not trivial to carry out the two tasks with limited wireless resources. Moreover, due to the high human cost and the environmental strain, frequent battery replacement for massive IoT is prohibitive. Therefore, energy, communication and computation are listed as three critical issues of B5G cellular IoT. Then, we introduce the key techniques to address these three crucial factors, i.e., wireless power transfer, non-orthogonal multiple access and over-the-air computation.

In Chap. 2, we study the issue of convergence of energy and communication in B5G cellular IoT with a massive number of terminal devices. In particular, we consider a practical scenario of a sustainable B5G cellular IoT enabled by SWIPT, where the IoT devices have a non-linear EH receiver and perform imperfect SIC due to a limited capability. To realize efficient convergence of energy and communication in B5G cellular IoT, the key is to effectively coordinate the co-channel interference due to non-orthogonal transmission. This is because co-channel interference has two sides of effects on the performance of B5G cellular IoT. On the one hand, co-channel interference decreases the quality of received signals for ID. On the other hand, co-channel interference increases the amount of received signals for EH. In general, spatial beamforming and power allocation are utilized to coordinate the co-channel interference. It is well known that the availability of CSI at the BS is the key to perform spatial interference coordination. Without loss of generality, we design the B5G cellular IoT network with three different CSI models including full CSI, imperfect CSI with channel quantization error bounded in an ellipsoid, and imperfect CSI with channel estimation error modeled by Gaussian stochastic process. Corresponding optimization algorithms are proposed to effectively alleviate the impacts of adverse factors as well as improve the overall performance. Extensive simulation results are presented to validate the effectiveness of the proposed algorithms.

In Chap. 3, we focus on the issue of convergence of energy and computation in B5G cellular IoT. Especially, we consider a computation-centric B5G cellular IoT network operated in the TDD mode, where a multi-antennas BS plays two roles, i.e., a power beacon in the downlink and a data fusion center in the uplink for multi-modal IoT devices equipped with multiple antennas. At first, the BS utilizes the WPT technique to charge IoT UEs via energy beamforming. Then, all IoT devices transmit a set of multi-modal data to the BS simultaneously with the harvested energy. Enabled by Aircomp, the BS designs a computation receiver to recover the targeted signal directly. Moreover, AirComp in B5G cellular IoT can combine

MIMO techniques, namely MIMO AirComp, to spatially multiplex multi-function computation by exploiting spatial degrees of freedom provided by large-scale antenna arrays, and can further reduce computation errors by using spatial beamforming. Hence, the key of designing of B5G cellular IoT lies in beamforming optimization. It is known that beamforming design is closely linked to the CSI. However, in B5G cellular IoT with massive access, it is only able to obtain partial or even no CSI. In other words, it is necessary to take the uncertainty of CSI into consideration for beamforming design, namely robust beamforming. In this context, in order to realize efficient convergence of energy supply and data aggregation in B5G cellular IoT, a robust design algorithm is provided by jointly optimizing beamforming of both WPT and AirComp. Simulation results validate the robustness and effectiveness of the proposed algorithm over the baseline ones.

In Chap. 4, we investigate the issue of convergence of communication and computation in B5G cellular IoT, where each IoT device has two independent signals, one for computation, and the other for communication. For communication, highly accurate sensing information at IoT devices are sent to the BS through by using non-orthogonal communication over wireless multiple access channels. Meanwhile, for computation, AirComp is adopted to substantially reduce latency of massive data aggregation via exploiting the superposition property of wireless multiple-access channels. Specifically, each IoT device carries out beamforming for coordinating the communication signal and the computation signal to be transmitted respectively, and sends a superposition coded signal to the BS over the uplink channel. On the one hand, enabled by the AirComp technique, the BS receives the computation results directly via concurrent data transmission without recovering individual data, and then utilizes a computation receiver to obtain the targeted function signal. On the other hand, the BS decodes the sensing signals of each device through communication receivers. To achieve effective integration of computation and communication under practical but adverse conditions, a robust algorithm is proposed by jointly optimizing transmit power and receive beamforming, with the goal of minimizing the computation error of computation signals while guaranteeing the requirement of communication signals. Extensive simulation results show that the proposed robust algorithm can realize efficient convergence of communication and computation with limited wireless resources.

In Chap. 5, we concentrate on a sustainable B5G cellular IoT integrating energy, communication and computation, where a BS equipped with a large-scale antenna array serves a massive number of multiple-antennas IoT devices. Note that IoT devices equipped with sensors collect information from the surroundings or human, and then transmit it to the BS for further decoding, computation, or analysis. Thus, the signals from the IoT devices have two functions, one for data aggregation based on multiple devices' signals, and the other for information transmission based on individual device's signal. However, it is not trivial to carry out the two tasks with limited wireless resources. In this case, NOMA is applied into cellular IoT to realize seamless access of a massive number of devices. Meanwhile, MIMO AirComp is used to reduce latency of massive data aggregation by exploiting the superposition property of wireless multiple-access channels. To effectively address the critical issues of B5G cellular IoT, i.e., energy supply, data aggregation and information transmission, we design a comprehensive framework. Firstly, the BS charges massive IoT devices simultaneously via the WPT technique in the downlink. Secondly, IoT devices with harvested energy carry out the computation task and the communication task in the uplink via AirComp and non-orthogonal transmission over the same spectrum. To improve the overall performance of energy, communication and computation, we propose a joint beamforming design algorithm for the BS and the IoT devices with the goal of minimizing the computation distortion, while guaranteeing the SINR requirements of communication signals. Simulation results validate the effectiveness of the proposed algorithm in B5G cellular IoT.

6.2 Future Works

Although theory and technique of the convergence of energy, communication and computation in B5G cellular IoT have been studied, there still are many challenging issues to be addressed for the design of a large-scale edge-intelligent B5G wireless network. In the following, we list some initial ideas and research directions in future works.

1. In B5G cellular IoT, there are a massive number of IoT devices connected to wireless networks for automating the operations of our daily work and life, thus providing intelligent services [1, 2]. In this context, one critical challenge is the need of ultra-fast wireless data aggregation [3]. In this book, we adopt a novel computation framework, namely AirComp, to reduce the latency and improve the spectrum efficiency. As the computation gets more complicated, more advanced big data technologies such as machine learning are required [4]. In particular, since the rapid growth in storage capacity and computational power of terminal devices, federated learning as a promising on-device distributed machine learning solution makes it possible for IoT devices to process data locally instead of risking privacy by sending data to the cloud or networks [5]. Despite the benefits of low latency, low cost, and high privacy of federated learning, communication bandwidth remains a bottleneck for globally aggregating the locally computed updates [6]. Thus, it is desired to design high communication-efficient schemes of federated learning according to the characteristics of B5G cellular IoT.

2. B5G cellular IoT networks with high edge-intelligence are expected to meet the requirements of ultra-low latency, ultra-high efficiency, ultra-high reliability and ultra-high density connectivity, which enables various envisioned IoT applications, such as smart city, smart manufacturing, smart transportation, e-health care, etc. [7, 8]. This book aims to address three critical problems in B5G cellular IoT, namely energy, communication and computation. In fact, other than theses issues, sensing is an important part of cellular IoT. For example, in smart cities, a large number of IoT devices are mainly used for environmental sensing [9]. As a result, there are a massive column of sensing information that has to be transferred from

IoT devices to the BS [10]. However, it is not a trivial task to transfer highly accurate sensing information over limited radio spectrum. Moreover, how to realize effective integration of sensing, communication and computation for supporting heterogeneous services is also an extremely challenging issue in B5G cellular IoT.

3. In general, B5G cellular IoT has the characteristics of low power, massive connectivity, and wide coverage. To support massive connectivity over limited radio spectrum, IoT devices should share the same spectrum [11]. As a result, massive access is susceptible to eavesdropping owing to the broadcast nature of wireless channels [12, 13]. With the fast development of information techniques, the eavesdropping capability of malicious nodes is increasingly strong, resulting in much more complicated cryptography techniques. Because most of IoT devices are simple nodes with limited computational capability, the complexity of cryptography techniques might be unaffordable. In this context, as a compliment of cryptography techniques, physical layer security techniques are applied into B5G cellular IoT [14, 15]. However, in B5G cellular IoT with massive connections, the received signal may suffer from severe co-channel interference, which degrades the performance of physical layer security [16]. Thus, it is necessary to design physical layer security techniques based on the features of B5G cellular IoT.

4. Cellular IoT is a key component of B5G wireless networks. Thus, it makes sense to apply the B5G NR techniques to further unlock the potential of the cellular IoT [17]. In this book, we have adopted several effective techniques, like massive MIMO and NOMA, to improve the overall system performance. In fact, there are lots of new promising techniques which are suitable for cellular IoT. For example, millimeter wave (mmWave) can provide a huge radio spectrum for short-distance communications [18, 19]. Moreover, intelligent reflecting surface (IRS) can be used to boost the received signal power, thus improving the achievable performance [20, 21]. Besides, the new waveform techniques, e.g., filter bank multi-carrier (FBMC), universal filtered multi-carrier (UFMC), generalized frequency division multiplexing (GFDM), and filtered OFDM (F-OFDM) can significantly enhance the efficiency, reliability, and flexibility of cellular IoT [22, 23].

5. This book provides several effective solutions for convergence of energy, communication and computation in B5G cellular IoT. In particular, resource allocation and beamforming design are optimized according to instantaneous CSI with the assumption of fixed and low-mobility cellular IoT. However, for the high-mobility IoT devices, e.g., UAV and vehicular equipments, the associated channels vary very fast, resulting in a short channel coherent time [24, 25]. In other words, it is difficult for the BS to obtain instantaneous CSI. Fortunately, since statistical CSI, i.e., channel mean and variance, remains constant within a relatively long time. Moreover, statistical CSI can be easily obtained at the BS by averaging over channel realizations [26]. Hence, it is desired to design the convergence of energy, communication and computation based on statistical CSI in the high-mobility cellular IoT scenarios.

References

1. Chen X, Ng DWK, Yu W, Larsson EG, Al-Dhahir N, Schober R (2020) Massive access for 5G and beyond. IEEE J Sel Area Commun (99):1–24
2. Al-Fuqaha A, Guizani M, Mohammadi M, Aledhari M, Ayyash M (2015) Internet of Things: a survey on enabling technologies, protocols, and applications. IEEE Commun Surv Tutor 17(4):2347–2376
3. Verma S, Kawamoto Y, Fadlullah ZM, Nishiyama H, Kato N (2017) A survey on network methodologies for realtime analytics of massive IoT data and open research issues. IEEE Commun Surv Tutor 19(3):1457–1477
4. Mohammadi M, Al-Fuqaha A, Sorour S, Guizani M (2018) Deep learning for IoT big data and streaming analytics: a survey. IEEE Commun Surv Tutor 20(4):2923–2960
5. Wang S, Tuor T, Salonidis T, Leung KK, Makaya C, He T, Chan K (2019) Adaptive federated learning in resource constrained edge computing systems. IEEE J Sel Areas Commun 37(6):1205–1221
6. Chen Y, Sun X, Jin Y. Communication-efficient federated deep learning with layerwise asynchronous model update and temporally weighted aggregation. IEEE Trans Neural Net Learn Sys. https://doi.org/10.1109/COMST.2015.2444095
7. Chen X (2019) Massive access for cellular internet of things theory and technique. Springer, Germany
8. Dawy Z, Saad W, Ghosh A, Andrews JG, Yaacoub E (2017) Toward massive machine type cellular communications. IEEE Wirel Commun 24(1):120–128
9. Wu H, Zhang Z, Jiao C, Li C, Quek TQS (2019) Learn to sense: a meta-learning-based sensing and fusion framework for wireless sensor networks. IEEE Internet Things J 6(5):8215–8227
10. Perera C, Talagala DS, Liu CH, Estrella JC (2015) Energy-efficient location and activity-aware on-demand mobile distributed densing platform for sensing as a service in IoT clouds. IEEE Trans Comput Soc Sys 2(4):171–181
11. Shirvanimoghaddam M, Dohler M, Johnson SJ (2017) Massive non-orthogonal multiple access for cellular IoT: potentials and limitations. IEEE Commun Mag 55(9):55–61
12. Granjal J, Monteiro E, Silva JS (2015) Security for the internet of things: a survey of existing protocols and open research issues. IEEE Commun Surv Tutor 17(3):1294–1312
13. Keoh SL, Kumar SS, Tschofenig H (2014) Securing the internet of things: a standardization perspective. IEEE Internet Things J 1(3):265–275
14. Chen X, Ng DWK, Gerstacker W, Chen H-H (2017) A survey on multiple-antenna techniques for physical layer security. IEEE Commun Surv Tutor 19(2):1027–1053
15. Wang N, Wang P, Alipour-Fanid A, Jiao L, Zeng K (2019) Physical-layer security of 5G wireless networks for IoT: challenges and opportunities. IEEE Internet Things J 6(5):8169–8181
16. Qi Q, Chen X, Zhong C, Zhang Z (2020) Physical layer security for massive access in cellular internet of things. Sci China Inform Sci 63(2):1–12
17. Wong VWS, Schober R, Ng DWK, Wang L-C (2017) Key technologies for 5G wireless systems. Cambridge University Press, Cambridge, UK
18. Rappaport TS, Sun S, Mayzus R, Zhao H, Azar Y, Wang K, Wong GN, Schulz JK, Samimi M, Gutierrez F (2013) Millimeter wave mobile communications for 5G cellular: it will work!. IEEE Access 1:335–349
19. Roh W, Seol J-Y, Park J, Lee B, Lee J, Kim Y, Cho J, Cheun K, Aryanfar F (2014) Millimeter-wave beamforming as an enabling technology for 5G cellular communications: theoretical feasibility and prototype results. IEEE Commun Mag 52(2):106–113
20. Hu S, Rusek F, Edfors O (2018) Beyond massive MIMO: the potential of data transmission with large intelligent surfaces. IEEE Trans Sig Process 66(10):2746–2758
21. Yu G, Chen X, Zhong C, Lin H, Zhang Z (2020) Large intelligent reflecting surface enhanced massive access for B5G cellular internet of things. In: Proceeding of IEEE VTC, Antwerp, Belgium, pp 1–6
22. Lien S-Y, Shieh S-L, Huang Y, Su B, Hsu Y-L, Wei H-Y (2017) 5G new radio: waveform, frame structure, multiple access, and initial access. IEEE Commun Mag 55(6):64–71

23. Farhang-Boroujeny B, Moradi H (2016) sOFDM inspired waveforms for 5G. IEEE Commun Surv Tutor 18(4):2474–2492
24. Chakareski J (2019) UAV-IoT for next generation virtual reality. IEEE Trans Image Process 28(12):5977–5990
25. Zhang Q, Jiang M, Feng Z, Li W, Zhang W, Pan M (2019) IoT enabled UAV: network architecture and routing algorithm. IEEE Internet Things J 6(2):3727–3742
26. Choi J (2016) On the power allocation for MIMO-NOMA systems with layered transmission. IEEE Trans Wirel Commun 15(5):3226–3237